AF694969

SUSTAINABLE DEVELOPMENT

Volume 2

Mountain World in Danger

Climate change in the forests and mountains of Europe

Full list of titles in the set
Sustainable Development

Volume 1: Global Environment Outlook 2000
Volume 2: Mountain World in Danger
Volume 3: Vanishing Borders
Volume 4: Atlas of Nepal in the Modern World
Volume 5: Caring for the Earth
Volume 6: Community and Sustainable Development
Volume 7: One World for One Earth
Volume 8: Strategies for National Sustainable Development
Volume 9: Strategies for Sustainability: Africa
Volume 10: Strategies for Sustainability: Asia
Volume 11: Strategies for Sustainability: Latin America
Volume 12: Wasted
Volume 13: Dam the Rivers, Damn the People
Volume 14: Rising Seas
Volume 15: Tomorrow's World
Volume 16: Social Change and Conservation
Volume 17: Threats Without Enemies
Volume 18: Progress for a Small Planet
Volume 19: Women and the Environment in the Third World
Volume 20: World at the Crossroads

Mountain World in Danger

Climate change in the forests and mountains of Europe

Sten Nilsson and David Pitt

earthscan
publishing for a sustainable future

London • New York

First published in 1990

This edition first published in 2009 by Earthscan

Copyright © Barbara J Cummings, 1990

All rights reserved

ISBN 978-1-84407-933-9 (hardback)
ISBN 978-1-84407-931-5 (Sustainable Development set)
ISBN 978-1-84407-930-8 (Earthscan Library Collection)
ISBN 978-0-41584-963-0 (paperback)
For a full list of publications please contact:

Earthscan
2 Park Square, Milton Park, Abingdon, Oxon OX14 4RN
Simultaneously published in the USA and Canada by Earthscan
711 Third Avenue, New York, NY 10017
Earthscan is an imprint of the Taylor & Francis Group, an informa business

First issued in paperback 2013

Earthscan publishes in association with the International Institute for Environment and Development

A catalogue record for this book is available from the British Library

Library of Congress Cataloging-in-Publication Data has been applied for

Publisher's note
The publisher has made every effort to ensure the quality of this reprint, but points out that some imperfections in the original copies may be apparent.

At Earthscan we strive to minimize our environmental impacts and carbon footprint through reducing waste, recycling and offsetting our CO_2 emissions, including those created through publication of this book.

MOUNTAIN WORLD IN DANGER

Climate Change in the Mountains and Forests of Europe

Sten Nilsson and David Pitt

Earthscan Publications Limited, London
in association with
Alp Action and IIASA

First published 1991 by
Earthscan Publications Limited
3 Endsleigh Street, London WC1 0DD

Copyright © Alp Action, Bellerive Foundation and IIASA 1991

All rights reserved

British Library Cataloguing in Publication Data
Nilsson, Sten
Mountain world in danger: climate change in the forests and mountains of Europe
I. Title II. Pitt, David
363.73

ISBN 1-85383-118-2

Cover design by David King
Production by Bob Towell
Typeset by Bookman Ltd, Bristol

Earthscan Publications Ltd is an editorially independent subsidiary of the International Institute for Environment and Development (IIED; charity number 800066).

Contents

List of Figures and Tables

1: Climate Change: Facts and Theories

2: Mountain and Cryosphere Situations

3: Climate Change and Mountain Regions

4: Climate Change and Mountain Forests

5: What Can Be Done?

6: International Responses

Abbreviations

ETH	Federal Institute of Technology, Zurich
GCM	Global Circulation Model
GDFL	Geophysical Fluid Dynamics Laboratory
GIPRI	Institut international de recherches pour la paix, Geneva
GISS	Goddard Institute for Space Studies
ICSU	International Council of Scientific Unions
IIASA	International Institute for Applied Systems Analysis
INSTRA	International Institute for Strategic Studies on the Environment Foundation, Stockholm
IPCC	Intergovernmental Panel on Climate Change
IUCN	World Conservation Union
NGO	Non-Governmental Organization
OSU	Oregon State University
RSWG	Response Strategy Working Group
SCAR	Scientific Committee for Antarctic Research (ICSU)
SCOPE	Scientific Committee on Problems of the Environment (ICSU)
UKMO	United Kingdom Meteorological Office
UN	United Nations
UNCED	United Nations Conference on Environment and Development
UNCTAD	United Nations Conference on Trade and Development
UNEP	United Nations Environment Programme
UNESCO	United Nations Educational, Scientific and Cultural Organization
WHO	World Health Organization
WMO	World Meteorological Organization
WWF	World Wide Fund for Nature (previously World Wildlife Fund)

Preface

Climate change, especially the greenhouse effect, has emerged as perhaps the greatest problem facing the world at the end of the twentieth century. Rarely before has there been such a consensus amongst scientists and NGOs that something needs to be done, even if all the research has not been completed.

In the great climate debates, the mountains have not often been mentioned, although they occupy 40 per cent of the world's surface. Mountains and the cryosphere (the frozen regions) play a major part in the climate and hydrological system, especially reflecting heat back to space and acting as a reservoir of vital water. The Alp Action programme has pushed mountains up the political agenda. This short book summarizes for the first time the possible effects of warming, particularly on the European mountains. The evidence is frankly alarming, the more so since it is presented in an unemotional way. The conclusions deserve careful consideration at the highest levels, for the implications go far beyond the mountains.

Clearly something must be done urgently, and on a global scale. This action can start with all of us, who should seek ways of conserving energy. There is a need for a co-operative effort amongst all the organizations and individuals concerned with the problem. There is also a need for dialogue – not confrontation – between those inside and outside politics. The publicity given to climate change has created a unique opportunity for raising awareness and stimulating action. The mechanisms for change exist in a world where there is a new atmosphere of international co-operation. The negotiations for a Climate Convention are a splendid opportunity. The forthcoming World Conference on Environment and Development will, it is hoped, provide a forum and a stimulus for a new environmental order in which mountains will receive appropriate attention.

Alp Action and the Bellerive Foundation are pleased that its network of outstanding scientists could produce this book and hope that it will be useful for briefing, education and reference.

Equally significant is the fact that this project was itself the outcome of an alliance between industry, scientists and ecologists. Special thanks are due to Unigestion for their support, without which the project could not have been attempted. We owe a debt of gratitude to The Royal Swedish Academy of Forestry and Agriculture and the International Institute for Applied Systems Analysis (Austria), who worked with us on the scientific analysis. I look forward to this network continuing to put before the public and decision-makers sensible, authoritative information for vital action.

Prince Sadruddin Aga Khan
President, Bellerive Foundation, Switzerland

Foreword

Air pollution in different forms has been described as the most serious threat to our environment. Lakes and soils have become acidified. Forests are damaged over vast areas. Serious warnings are given about climate changes and the greenhouse effect. We are dealing with complex chemical stress factors causing widespread disturbances in ecosystems. There is concern that a continuation of recent trends will lead to most undesirable consequences.

The problem is international, even if each local situation is different. Strategies to combat undesirable impacts must be set in the international arena because air pollution crosses boundaries, as does trade in raw materials for processing.

Intensive and special attention is being paid to the increasing concentrations of gases such as carbon dioxide, methane, freons and nitrous oxide in the global atmosphere. But studies are not enough; there must be long-term energy policies in which the control of air pollution and environmental protection plays a most important role. Decision-makers should adapt to the risks inherent in the management of environmental issues. The public must be educated about the consequences of climate and inappropriate behaviour.

The purpose of this book is to produce a view of possible future interrelationships between climate change and forests in the high mountains of Europe, the complexities and the necessary strategies and actions in response to changes. Even if awareness and knowledge about the possible negative effects of air pollution have improved in recent years, the messages of this book should be inserted into the process of global environment protection. We have to establish an efficient and effective exchange of knowledge between politics, industry and science. This is surely the common responsibility in international co-operation.

Lennart Schotte
President, The Royal Swedish Academy of Forestry and Agriculture
Chairman, Board of INSTRA Foundation

Acknowledgements

This work is the product of many hands, and it would be impossible to thank individually all those who helped – especially in Alp Action, the Bellerive Foundation, the International Institute for Strategic Studies on the Environment Foundation, Stockholm (the original initiators of the project), the International Institute for Applied Systems Analysis, the Royal Swedish Academy of Forestry and Agriculture, UNEP, WMO and the Intergovernmental Programme on Climate Change. But some special thanks are in order.

Professor Lennart Schotte of the Royal Swedish Academy of Forestry and Agriculture kindly reviewed the text. Marc Sursock of Alp Action, Bellerive provided guidance for the editing and production, whilst Javed Ahmad, also of Alp Action, designed the cover. M. Sundararaman (WMO) facilitated contacts with, and materials from, the IPCC. Frances Baron (UNEP) was, as always, a mine of information and leads. Roger Newson (WMO) presented materials on the ALPEX project. Professor Ozenda of Grenoble, the doyen of Alpine scholars, provided much insight into the contemporary problems. Martin Holdgate (IUCN) kindly supplied the interesting graphics from the Commonwealth Secretariat. Jeremy Leggett and Paul Hohnen of Greenpeace sent excellent materials and comments for us to use. Edouard Dommen (UNCTAD) provided interesting data on appropriate environmental economic models. Dr W. Haeberli and Dr Roger Barry at the ETH (Switzerland) and Dr Roger Street (Canada) supplied materials on the cryosphere. Lars Nordberg of UNECE was helpful at many stages of the project. Bernado Zentilli and the staff at the United Nations Conference on Environment and Development kindly reviewed the text. Professor Ivo Rens of the University of Geneva Faculty of Law and GIPRI, in many hours of discussion, instilled reason and wisdom into the judgements on climate change. Able typing and editorial assistance was provided by Valerie Antonietti, Beatrice Reynolds, Annie Guyon and Cynthia Ramirez. Sten Nilsson (IIASA) and David Pitt (Alp Action) co-ordinated the inputs and wrote the original drafts for

Chapters 4 and 5 and 1, 2, 3 and 6 respectively. Neil Middleton at Earthscan and Robert Duis (IIASA) made helpful comments and expedited the publishing process.

Finally, the work could not have been undertaken at all without the generous support of Unigestion, who funded the project, and have a special interest in mountain forests. Bernard Sabrier, the Chairman of the Board, Unigestion SA, has written:

> Our natural heritage, which will constitute the well-being of our children, is above all composed of air, water, the flora and the fauna. The forests unite all these elements; this is why we have chosen this theme.

Introduction

This book is not just for scientists and ecologists. It is also intended for decision-makers, politicians, diplomats, teachers, and for the interested public. All have tended to neglect the mountains on their back doorsteps in preference to what are presumed to be more pressing problems further afield. Not only have mountains been neglected, but also the effects of climate change upon them. The present expectancy is of major warming. If the predictions of the greenhouse theorists are correct, there will be little ice or snow in the Alps by the year 2100.

Mountains have always had a central part to play in Europe's history and culture. The prehistoric settlement of Europe and even the present-day distribution of the great Romance, Germanic and Slavic cultures was greatly influenced by mountain geography. People moved along rivers, through the gates between mountain chains, and sought areas that were neither too cold nor too heavily forested. In front of the retreating ice were areas of loess soils, drier, open heathlands and steppes, more easily cultivated or grazed with prehistoric techniques. As warmer conditions prevailed, mountain grasslands were used, at least in the summer, especially by herding cultures.

Europe's mountains – or at least the peaks – are part of the global cryosphere, those frozen icy, snow-covered regions of the earth, which can cover about 60 per cent of its surface (see Table 1.3). But Europe's mountains are much more than snow and ice now under threat in a warmer world. Their ecosystem embraces forests, meadows, rivers, fauna, flora and great cultures. The plains below depend on the mountains for water, energy, agricultural products and tourism. Already the European mountains are under extreme pressure, in part because of climate changes. Forests are dying in drier polluted conditions. The hydrological system is disturbed. Fauna and flora are threatened. The tourist industry is suffering. These problems will certainly intensify in the future.

This book sets out to provide a succinct summary of probable future scenarios for the mountains and forests of Europe in times of climate change, and to outline the responses necessary in this situation. We do not propose a detailed scientific treatise but rather a commentary on the political processes surrounding the use and abuse of the atmosphere. We will not waste time on complex definitions. The European mountains are simply defined as the elevated regions from the Atlantic to the Urals, from the Arctic to the Mediterranean. Europe's mountains and forests cannot, however, be discussed in isolation from the global biosphere of which they are part. Therefore, there is discussion of those factors in the wider natural world and in society which affect what is happening in Europe. It may be that the most important solutions to the ecological problems of mountain and forest ecosystems are to be found in faraway centres of political power and in the behaviour of ordinary consumers. And while the book focuses on Europe, there are lessons here for other parts of the mountain world where the problems of industrialization are on the horizon.

Our book looks to the future rather than the past, but it is not intended to be a cookbook of guaranteed recipes for success. At present, there is great uncertainty about the future of ecological and climatic changes. This uncertainty is especially true of mountain regions. Trends of warm winters may suddenly give way to extreme cold. Some have argued that this uncertainty precludes drastic – or, indeed, any – action. We feel, however, that the uncertainty presented in the different scenarios and stories demands, on the contrary, precautionary, preventive and protective action.

The data come from several different sources. There are weather records and vegetation indicators. Use is made here of the Climate Data Base of IIASA (International Institute for Applied Systems Analysis, Austria), one of the partners in the project. Secondly, extensive use has been made of the materials collected by the IPCC (Intergovernmental Panel on Climate Change, Geneva).

We begin by reviewing current models about climate change. Most attention is paid to the greenhouse warming theory, though there is also discussion of cooler models, and steady-state ideas, such as Gaia. The "business as usual" model (i.e. assuming more or less current levels of energy use) is discussed in depth because it has been the focus of most scientific interest and also reflects political expectations and plans. The dynamic mountain situation, where data and prediction are uncertain, complicates, confuses and confounds the leading models. We discuss this in Chapter 2

The third chapter spells out the situation assuming the "business as usual" scenario in the mountains. There is a detailed discussion of the range of physical and socioeconomic effects, some catastrophic and most alarming. Special attention is paid to the effects on forests in Chapter 4.

The book is not, however, just a catalogue of doom and gloom. A chapter is devoted to the options that are available and the best possible strategies. Chapter 6 spells out the international responses being taken or contemplated. The book concludes with a set of next steps. In the Appendices, key documents relating to climate change in the mountains are reproduced.

Throughout there is a running argument that mountains are an essential – perhaps the most essential – and most threatened ecosystem vital to the world climate, water supply, vegetation, wildlife and culture.

Chapter 1

Climate Change: Facts and Theories

At present, there are a number of apparently conflicting theories about what is happening to the climate both globally and in specific regions. Two broad groups of theory may be identified: in the first, the catastrophe group, rapid irreversible changes are perceived; the second tends to see less a supercrisis than a pseudocrisis: cyclical or harmonic processes in which there are winners and losers in different regions and a measure of stability.

We may start with the best-publicized catastrophe theory, the greenhouse, which Gribbin and others have called the "heat trap".[1] This means leaving aside the interesting, if somewhat speculative, histories which talk of devastating meteorites or earth wobbles, though the history of climate change has undoubtedly included such catastrophic events.

The most widely accepted theory of climate-change causation is the greenhouse effect, even if there are variants on the theme. The leading journal *Climate Change* polled scientists and found that the great majority believed in it.[2] The Intergovernmental Panel on Climate Charge (IPCC), convened by the World Meteorological Organization and the United Nations Environmental Programme, and the Second World Climate Conference agreed that there is a "natural greenhouse effect, which already keeps the earth warmer than it would otherwise be". Emissions from human activities substantially increase the atmospheric concentrations of the greenhouse gases – carbon dioxide, methane, chloroflurocarbons (CFCs) and nitrous oxide. These increases "enhance" the natural greenhouse, resulting in additional warming to the earth's surface.[3]

In the IPCC model of the greenhouse effect, some gases are transparent to incoming short-wave radiation from the sun but do not allow it to escape back to space. So the lower atmosphere is warmed. Table 1.1 summarizes data on the main greenhouse gases in the IPCC model, and their sources, though other authorities give different figures and interpretations.[4] The burning of fossil fuels may contribute as much as 90 per cent of the increase in carbon emissions.

Also acting as a windowpane is water vapour which, according to Lindzen,[5] is by far the most influential greenhouse gas (97 per cent). Unfortunately the critical water vapour occurs over 6 kilometres high, where data are scanty and unreliable. Water vapour may cool rather than heat if there is condensation into clouds.

Without the greenhouse, the earth would be an icy waste with a temperature of –30°C. When the shutters are too firmly closed, the temperature can rise to hundreds of degrees. The earth is literally between the chill of Mars and the inferno of Venus. Not everybody, however, agrees with the "consensus". One recent criticism of the IPCC claimed that the conclusions were "hot air", a political device to sound ominous when the scientific data were quite uncertain.[6] Others claim that those who "pooh-pooh" the greenhouse theory are in the pockets of those who want industrial expansion.

The debate is also concerned with how much temperature rises are part of natural cycles. The observational evidence is certainly very recent, basically starting in the mid nineteenth century though records go back more than three hundred years in England. It is also incomplete. Some of those records do show some evidence of natural cycles (for example, twenty years) though not everywhere, not the same cycles, not all the time.[7] These cycles have been explained by the moon (tidal variations) or solar activity. There are eleven-year cycles of sunspot activity and changes in solar magnetic fields after each cycle, as well as longer-term rhythms (forty years) discovered in Carbon 14 in tree rings. There is some association between low solar activity and cold conditions – as in the Little Ice Age in the seventeenth century – or with magnetic fields – as in the unexplained reversal of the earth's field after the last Great Ice Age. As sunspot activity increases there is more energy, but the contribution to radiative forcing (warming) on the earth is small. Politically influential (and funded) institutions, such as the Marshall Institute in Washington, have argued for such natural variation and ridiculed the greenhouse theorists. But nearly all the world's climatologists and many other scientists working on the IPCC see the present situation as unprecedented.[8]

Table 1.1: Greenhouse gases and human sources

	Carbon Dioxide (CO_2)	Methane (CH_4)	Nitrous oxide (N_2O)	Chloro-fluorocar-bons (CFCs)	Tropo-spheric Ozone
% contribution to "greenhouse effect" over period 1950–85	56	14	7	23	a
Concentration of greenhouse gases – pre-industrial (b)	275 ppmv	700 ppbv	280 ppbv	zero	15 ppbv
Concentration in 1988 (b)	350 ppmv	1700 ppbv	310 ppbv	0.26 ppbv (CFC-11) 0.44 ppbv (CFC-12)	335 ppbv
Annual growth of concentrations in 1980s	0.5%	0.5%	0.25%	5–5.5%	1%
Sources of greenhouse gases	Fossil fuel burning Deforestation / land – use changes	Rice paddy cultivation Rearing of ruminants (e.g. cows) Biomass burning Fossil fuel extraction and burning	Fertilizers Fossil fuel and biomass burning Land conversion for agriculture	Manufactured for solvents; aerosol spray propellants; foam packaging	Product of sunshine and pollutants: carbon monoxide; methane; other hydrocarbons nitrogen oxides

Notes: a. contributions of ozone not estimated, perhaps around 8 per cent of total.
b. ppmv is parts per million; ppbv is part per billion.

Source: Commonwealth Secretariat (1989).

The greenhouse effect, however, is very complicated, with a wide range of feedbacks with positive or negative effects, not all of which can be measured in computerized General Circulation Models (GCMs). Feedbacks "feed" on each other, producing effects which amplify warming or cooling. Such feedbacks may accelerate catastrophically.

Human factors play an important role in the creation of the greenhouse effect. The clearance of the tropical forests is associated with increases in human numbers in poor countries but also with the ever-increasing consumer demands in the industrialized world for both wood and hamburgers from grazing beef on cleared land. The average American consumes forty times as much energy as the average Ethiopian. The world car park numbers 400 million and is doubling every fifteen years, so that there will soon be more cars than poor people. Vehicle emissions are said to contribute over a third of the greenhouse gases, despite more stringent controls.

Pollution poisons as well as heats the biosphere, most significantly through an excess of acidity. Acid precipitations and soils are blamed for the sickness and premature death of 50 per cent of Europe's forest as well as for the disappearance of sensitive fauna and flora.

Some small products of industry have a major effect, the best known example being chlorofluorocarbons (CFCs) used as solvents, refrigerator fluids, aerosol propellants and in plastic foams. CFCs make up nearly a quarter of the greenhouse gases as well as destroying the ozone layer which protects the earth from damaging ultra-violet radiation. Figure 1.1 shows world production of CFCs and halons – mostly used in fire extinguishers – in 1985. An international protocol was signed in Montreal in 1987 with the objective of reducing CFC emissions.

The current emissions of methane (CH_4) are dominated by natural sources (wetlands, oceans and freshwaters). The estimated contribution from natural sources is 30 per cent of the total emissions. Each factor – rice production, fossil fuel production, domestic animals and biomass burning – is estimated to contribute by 13–16 per cent to the total emissions of CH_4. It is calculated that landfills make a contribution of roughly 8 per cent. Current emission sources of nitrous oxide emissions (N_2O) are predominantly natural too (75 per cent). Nitrogen fertilizers and biomass burning are thought to contribute by 5–9 per cent each in cultivated land and fossil fuels by 2–9 per cent each.

Figure 1.1: Production of CFCs and halons in 1985, expressed in metric tons

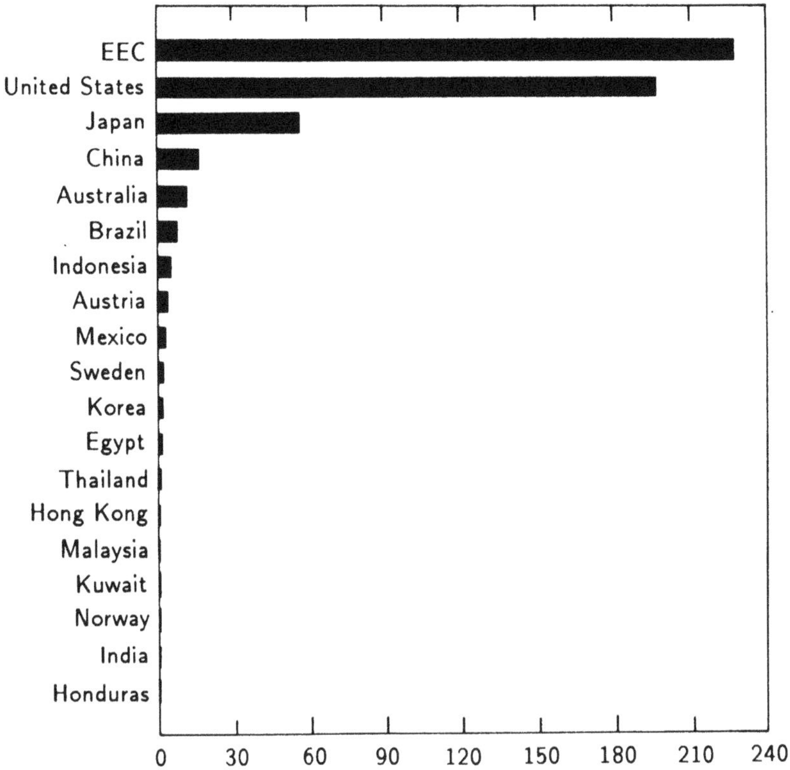

Source: US Environmental Protection Agency (1988)

The various gases have different relative influences on climate change. One concept to illustrate this is the "global warming potential" model developed by IPCC (1990). By using such an approach it is possible to convert greenhouse gas emissions to CO_2 equivalents. The global warming potentials and emissions by sector, region and gas are presented in Table 1.2 and Figure 1.2.

The human element, however, is recognized as only part of the reasons for global warming. After all, the planet has experienced warm interglacials before there was industrialization. Ice core and other analyses have shown them to be associated with increased CO_2 concentrations. An astronomical explanation of this phenomenon, the Milankovitch effect, has been widely accepted. Approximately every 90,000 years, the earth tilts slightly back and forth on its axis, allowing the ice sheet to creep gradually forward and then retreat. The CO_2 concentration has been explained by accidental forest-burning which may also create the cloudy conditions which favour cooling.

Table 1.2: Global warming potentials

Trace gas	GWP[a]
CO_2	1
CH_4	21
N_2O	21
CFC-11	3,500
CFC-12	7,300

[a]Global warming potentials following the instantaneous injection of 1Kg of each trace gas, relative to CO_2 for a 100-year time horizon.

Source: US Agency for International Development (1990).

Even today the human addition to the natural cycle may be quantitatively small, even if sometimes qualitatively significant. The human agency adds annually some 7 billion tonnes of carbon to the carbon flux compared to 200 billion tonnes exchanged between the atmosphere, living organisms and the ocean. There is a natural process of regulation and feedback. In the carbon flux, largely through photosynthesis (particularly forests and oceanic phytoplankton), there is a reduction in CO_2 and a conversion to oxygen.

However, CO_2 is increasing exponentially (Figure 1.3) along with temperature rises (Figure 1.4). The catastrophic forms of the greenhouse theory envisage a runaway effect, as presumably happened on the planet Venus. Here the atmosphere is almost entirely composed of CO_2 and the total greenhouse effect creates temperatures of above 500°C. As the sun warmed, it is presumed that the Venus water evaporated, disappearing into space and disintegrating into hydrogen and oxygen molecules through the strong action of the sunlight.

Figure 1.2 Relative contribution of 1985 emissions to radiative forcing

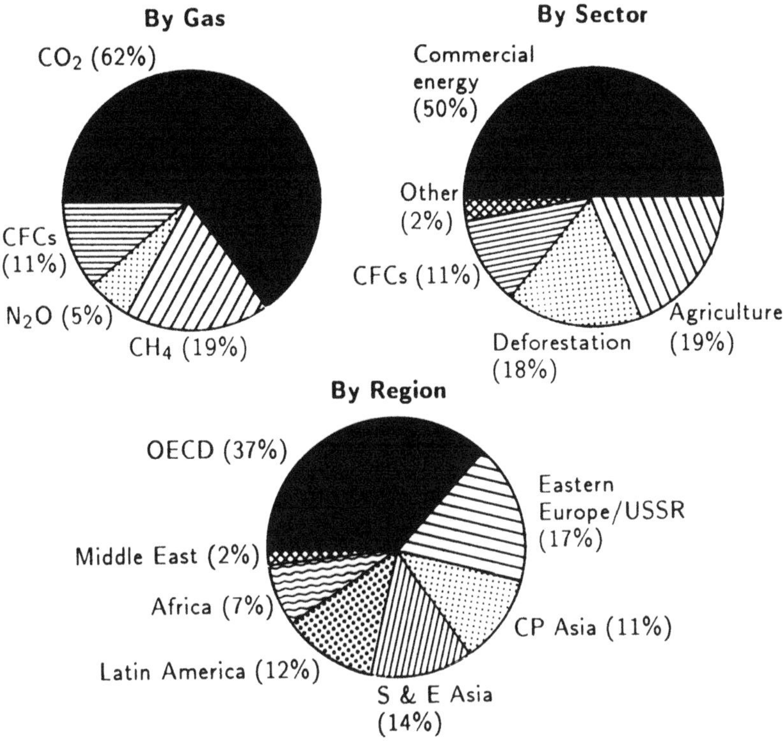

Source: US Agency for International Development (1990)

Similar feedbacks may take place in the water cycle. As there is warming, there is more evaporation and conversion to water vapour, which may amplify the greenhouse effect but also cool the planet if there is increased cloudiness and precipitation.

Until the 1980s, scientists thought the world would get colder for a variety of reasons[9] but mainly because of clouds from nuclear explosions, fires or volcanic eruptions. A contemporary minority ice age theory accepts certain elements of greenhouse causation but looks for different effects.[10] The focus is on the unseasonably cold winters and massive storms. The point is made that the first recorded snow fell on Miami in 1977, and since then such cold waves have become frequent in North America. The northern hemisphere snow cover has been extending, coming earlier or unseasonably, or staying longer. In time snow becomes ice through intermediate stages such as firn. Once snow lies for a season, melting is retarded. Even if the snow melts quickly, or comes only as frost, the crops may be destroyed – especially when, as often happens, tender shoots are affected. The increased snow cover reflects back sunlight (the albedo effect), so increasing cooling, while the mass of the snow weighs down on the earth, causing earthquakes and volcanoes (the toothpaste effect), which increase turbidity and coolness.

Figure 1.3: Record of the atmospheric concentration of carbon dioxide

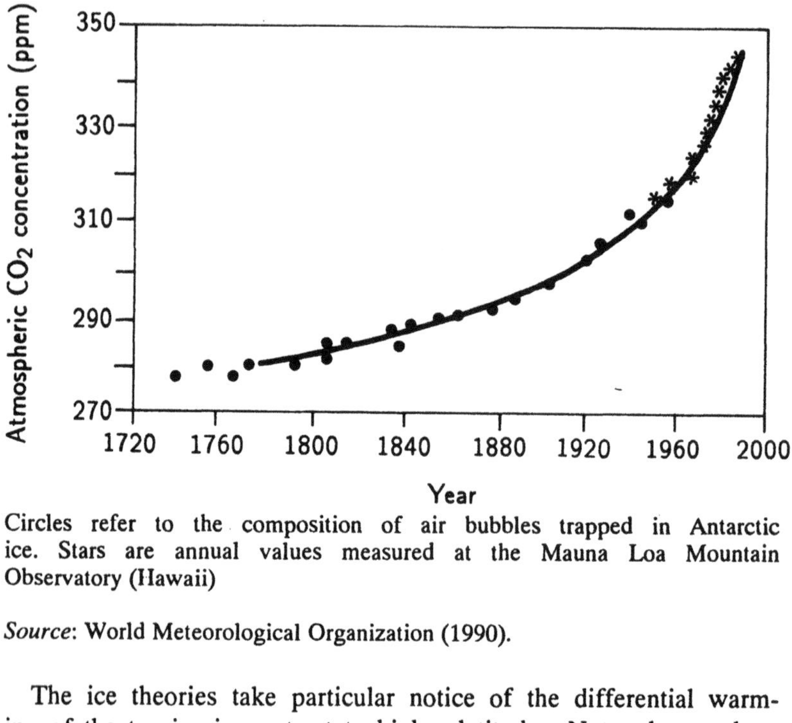

Circles refer to the composition of air bubbles trapped in Antarctic ice. Stars are annual values measured at the Mauna Loa Mountain Observatory (Hawaii)

Source: World Meteorological Organization (1990).

The ice theories take particular notice of the differential warming of the tropics in contrast to higher latitudes. Not only are there more storms as polar air rushes in towards the tropics but more cloudiness, which flows polewards with the circulation pattern, falling as increased precipitation, often as snow at high latitudes or altitudes. According to this theory, something like a 10 per cent sustained increase in cloud cover could bring on an ice age (see Figure 1.5).

Figure 1.4: Record of global mean air temperature change at the surface of the earth

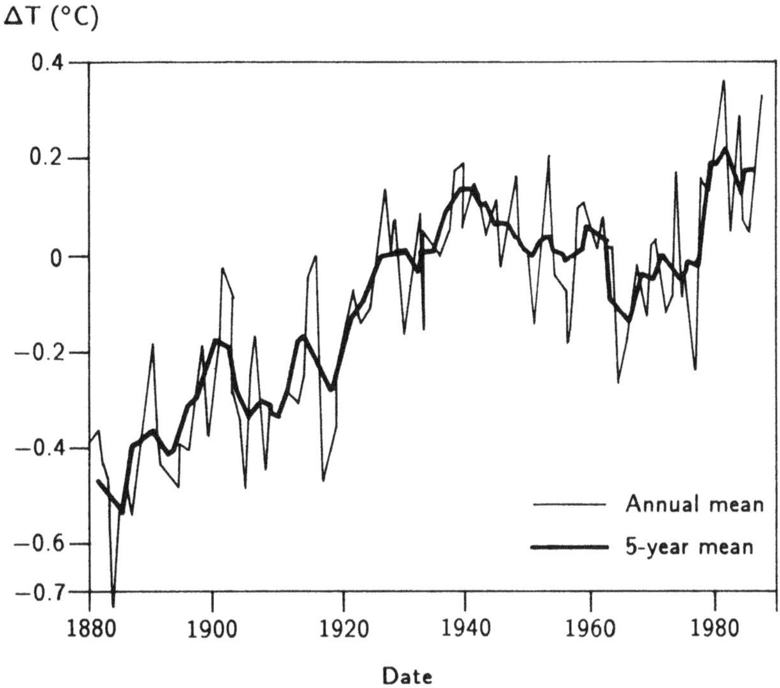

Source: World Meteorological Organization (1990).

There is, then, a "cold", not a "hot" greenhouse effect, which is compounded by the poor state of the forests, especially in temperate zones which cannot easily soak up CO_2. In Hameker's theory, forest death is explained by the demineralization of soils due to leaching and erosion. Forest death is therefore a cause rather than an effect of global warming. But the reversal of the theory does not stop there. The revitalization of the forests is triggered by the returning ice, which remineralizes the soils by crushing rocks, the dust of which is distributed as silt by rivers or in rainfall.

The ice age theory does at least feature the cryosphere, the snow and ice portions of the earth's surface. Even now, at what may be the warmest part of an interglacial, the cryosphere may extend to 60 per cent of the earth's surface in mid February (northern winter) (see Table 1.3).

Figure 1.5: Factors that may bring on the next Ice Age

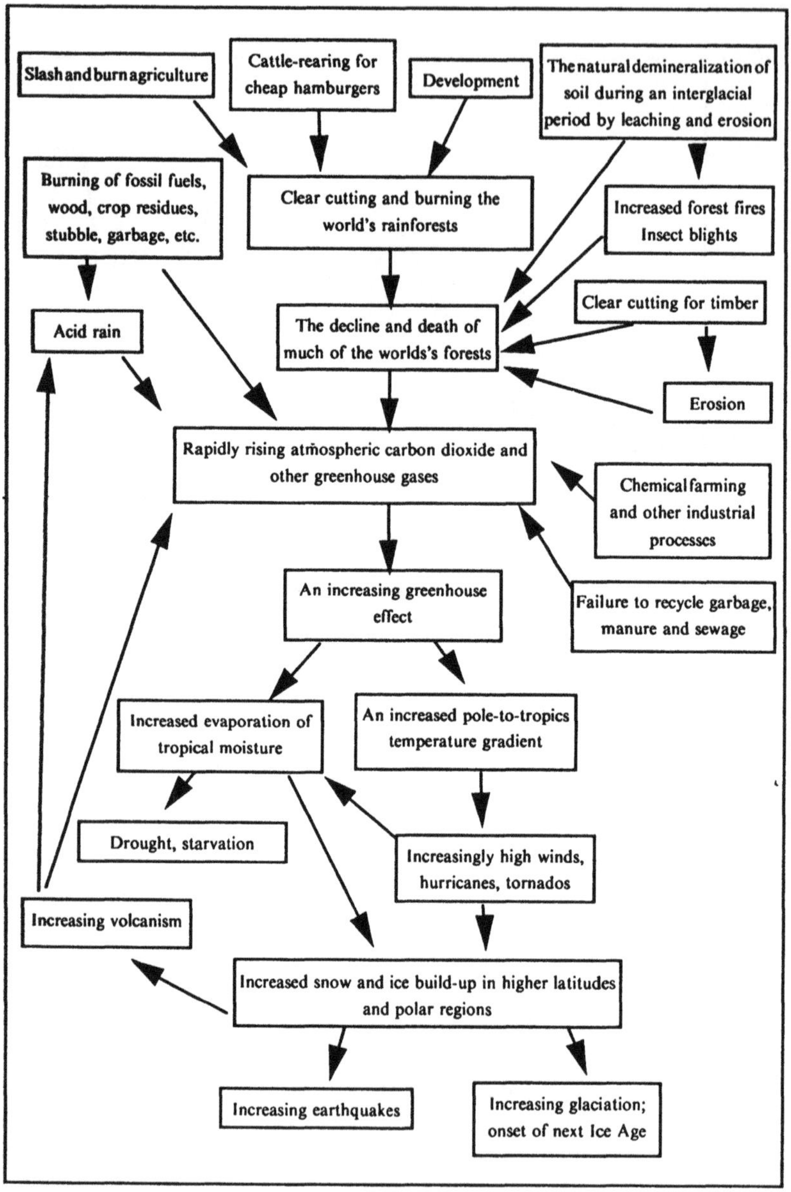

Source: Ephron, L. *The End* (Berkely: Celestial Arts, 1987).

Table 1.3: Components of the cryosphere

	Area ($10^6 km^2$)	**Ice volume ($10^6 km^2$)**	**Sea level equivalent (m)** *a*
Land ice: East Antarctica *b*	9.86	25.92	64.8
West Antarctica	2.34	3.40	8.5
Greenland	1.7	3.0	7.6
Small ice caps and mountain glaciers	0.54	0.12	0.3
Permafrost (excluding Antarctica):			
Continuous	7.6	0.03	0.08
		to	to
Discontinuous	17.3	0.07	0.17
Sea ice: Arctic *c*			
March	15.8	0.05	
September	7.6	0.02	
Antarctica *d*			
September	20.9	0.06	
February	4.0	0.10	
Land snow cover *e*			
N Hemisphere			
Early February	46.3	0.002	
Late August	3.7		
S Hemisphere			
Late July	0.85		
Early May	0.07		

a 400 000km^2 of ice is equivalent to 1m global sea level.

b Grounded ice sheet, excluding peripheral, floating ice shelves (which do not affect sea level). The shelves have a total area of 1.62 x 10^6km^2 and a volume of 0.79 x 10^6 km^2.

c Including the Sea of Okhotak and the Baltic Sea.

d Actual ice areas excluding open water are 18.4 and 3.6 million km^2.

e Snow cover includes that on land ice but excludes snow-covered sea ice.

Source: Barry, R. (personal communication).

Cloudiness (turbidity), the major factor in cooling, may be generated by a number of factors. In the 1970s, the work of the Max Planck Institute and elsewhere showed that even small nuclear explosions could create enough ashes in the troposphere to lower temperatures significantly. The USSR Academy of Sciences produced models which showed that after a nuclear war the earth might be totally blanketed by a cloud of soot within a month, and that this cloud might take a year to clear. These studies also concluded that such clouds might be formed from the firestorms generated by non-nuclear wars.[11]

Other significant sources of ash may follow volcanic explosions or forest fires. Indeed, one of the earliest analyses of climate change was carried out by Benjamin Franklin in the 1780s, when he was Ambassador of the newly independent United States in Paris. He suggested an association between the cold winters and cool summers and Icelandic volcanic activity.

The extent to which a nuclear winter, firestorm cloud, forest fires or volcanic eruptions may occur is not predictable, though UN and ICSU studies have talked of the possible risk of unprecedented consequences with temperature drops of 25–30°C.[12] Nuclear weapons have proliferated both horizontally and vertically, despite international treaties and nuclear-free zones. Every state and terrorist group may be able to possess the technology and all but the poorest groups may already have some capability.[13] The outcome of the Gulf crisis has proved that the threat of war is never far away, and the effects of the immense clouds produced by burning oil wells in Kuwait have yet to be measured. The Geneva International Peace Research Institute has recorded over three hundred wars since World War II. Nuclear power stations, which are vulnerable to either attacks or accidents, are still numerous despite Chernobyl and may increase as there is a move away from polluting fossil fuels. Forest fires are more common, associated with the drier conditions in some areas resulting from climate change, as well as the increased incidence of arson. Finally, unexpected events cannot be ruled out. *Nature* has reported that the comet that may have caused the death of the dinosaurs is coming back. The dust in its tail would be enough to trigger a major ice age.[14]

The sophisticated Gaia explanation which has stability rather than instability as a basic premise relies on biological analysis. Gaia is the newest phase of cosmological explanations. From antiquity the universe was explained theologically. After Newton the analogy was the machine. Today there is considerable interest in the notion of the universe as a living being. The focus is not on CO_2 gas but on oxygen, nitrogen, etc. – or rather, the synergy between organic and inorganic. For Lovelock the atmosphere is not a "mere catalogue of gases" but a living entity in which oxygen is the key influence, if not the most abundant gas (21 per cent compared to nitrogen's 79 per cent). Oxygen tension is the "electrical supply" for life as it oxidizes with other elements (redox potential). Lovelock compares the environment to a battery, with one electrode in the oxygen and the other in the food. Oxygen is produced from photolysis of water vapour in the upper layers of the atmosphere and the "burial" of fixed carbon from plants.[15]

With low levels of oxygen, as Pasteur showed, there can be only low levels of life, but too much oxygen is also very dangerous. If the oxygen levels were to reach 25 per cent, anaerobic life, which cannot live with oxygen, would decline. The world's forest could be destroyed by flash fires (as has happened in the earth's history). Anaerobes are significant in several ways – not only as the internal fauna and flora within living creatures but also in the muds and sediments on seabeds, marshes, wetlands and river estuaries. In both places, through bacterial action, the most important oxygen-regulating gas – methane – is produced. Nitrous oxide, like methane, is present only in very minute quantities in the atmosphere (one-third parts per million) and may regulate methane, oxygen and ozone. The Gaia argument is that there is a happy medium of ozone. The holes may be bad, but too much ozone may preclude ultraviolet light which is beneficial to human health – preventing rickets, for example. Another nitrogenous gas made in large quantities in the soil and seas is ammonia, which controls the acidity of the atmosphere.

By far the most prevalent gas in the atmosphere is nitrogen, which dilutes oxygen, maintaining sufficient air pressure or density for a stable climate. The atmosphere is seen as a kind of storehouse preventing the creation of nitrates which are toxic and, when dissolved in the sea, lead to excessive salinity inimical to life.

CO_2 is a minor gas, even if it is a major gas in the greenhouse panels, and it is absorbed not only through photosynthesis in trees and plants, but also non-photosynthetically by heterotrophic animals. The Gaia conclusion is that harmony is the natural state of the earth and that humankind must work as part of this harmony whereby temperature can be thermostatically controlled. The general prescription is to have more contact with nature, with the Gaia Mother Earth, and respect her logic. The Russian scientist Vernadsky, who provided much inspiration for Gaia, went further. For him the universe was dominated by a higher sphere than the biosphere – the noosphere – rather like the mind in the body. The noosphere allowed the beneficial intervention of the human intellect. Grinevald sees biospheric theories as a continuing thread in scientific thought since its origins.[16]

The most important recent models have been more pragmatic, generated by several hundred scientists and policymakers who have contributed to the IPCC. The IPCC models are important, drawing on more evidence than any other model, both from detailed (Euro-American) computer models and historical analogues (Soviet palaeoclimate). More significantly perhaps, IPCC models have already become the basis for official planning and policies, and have attracted a wide consensus.

Some of the virtues of the IPCC models, however, are also vices. The fact that models have been evolved through intergovernmental channels has increased the possibilities of political pressures. The fact that there is an overload of sometimes contradictory information has led to some shifting of the goal posts – with regard, for example, to temperature and sea-level predictions. The fact that so many scientists have been involved has increased bureaucratization, overlap and inappropriate sequences. The featuring of CO_2 in the greenhouse effect (in the political rather than the scientific statements) has been interpreted as a subtle pro-nuclear power argument.[17]

The IPCC process has been organized around four major working groups. The first assembled the scientific evidence, the second assessed the effects, the third suggested responses and the fourth considered the position of the Third World. The reports of the latter have depended on the scenarios suggested by the first. Some consternation was created when late in the exercise in May 1990, Working Group 1 (WG1) significantly lowered its estimates and made fuzzy the links between emissions, the greenhouse effect and climate changes. Since it was impossible to re-do many months of work and calculations, Working Group 2 at least retained the higher estimates.

WG1 is based on the uncertainty principle, even if absence of evidence is not necessarily evidence of absence. Only two things are felt to be certain: first, that there is a natural greenhouse effect which already keeps the earth warmer than it would otherwise be; secondly, that emissions resulting from human activities do substantially increase the atmosphere concentrations of greenhouse gases – CO_2, methane – the chlorofluorocarbons and nitrous oxide, enhancing the greenhouse effect and resulting in additional warming.

Beyond that, WG1 "calculated" with confidence that some gases are potentially more important than others in changing climate, and that this effectiveness can be estimated (e.g. CO_2 contributing about half the enhanced effect). Secondly, the atmosphere concentrations of the long-lived gases (CO_2, NO_x and CFCs) adjust only slowly to changes in emissions, so that high future concentrations are already assured. CO_2 notably lingers for two hundred years. To stabilize concentrations at today's levels, emissions from human sources would need to be reduced immediately by at least 60 per cent. A major factor is that even if emissions in developed countries may reduce, those in other parts of the world are still increasing (see Table 1.4). A string of uncertainties are, however, introduced, including the sources and sinks of greenhouse gases – all of which change judgements and introduce surprises.

Table 1.4: Sources of CO2 emissions from fossil fuels

	1950 (%)	**1980 (%)**
North America	44.7	26.7
Western Europe	23.4	16.5
USSR/Eastern Europe	18.0	24.2
Japan/Australia	2.8	5.8
China/Communist Asia	1.4	8.5
Developing	5.7	12.2
Others	3.9	6.0
	100	100
Total	**1.6 bn tonnes**	**5.17 bn tonnes**

Source: Commonwealth Secretariat (1989)

Taking all this into account, the WG1 model interprets the rise of global mean surface air temperature by 0.3°–0.6°C over the last hundred years and the five global average warmest years in the 1980s as being of the same magnitude as natural climate variability.

The unequivocal detection of the enhanced greenhouse effect from observations is not likely for a decade or more. The report sees less doom and gloom than a situation of winners and losers, and even that enhanced levels of CO_2 may increase the productivity and the efficiency of its use of water.

As a consequence, WG1 concludes that the average temperature rises in the next hundred years will certainly be more than those in the last century, or 1° with doubling CO_2 by 2025 (as compared to their own previous estimate of 3° with doubling CO_2). With controls on emissions, the total rise in temperature could be only 0.1°C per decade. Sea-level rises predictions are reduced to only 6 centimetres compared to 1–2 metres. None the less, this rise is seen as a major departure from a much slower historical experience and a real problem because of its rapidity.

The WG1 model was not, of course, pulled out of the air, but based on extensive observations and recent computer models (especially from the UKMO). These observations, however, even of simple parameters such as temperature and precipitation, do not add up to a total picture. Observation stations are unevenly distributed throughout the globe. There are many more in the northern than in the southern hemisphere, for example, and – more significantly – there are very few in the three-quarters of the world that is ocean or the half of the world that is mountains. Even more significantly, perhaps, most weather stations are found in towns. This skews temperature reporting because towns are part of "heat islands" which are significantly warmer than the surrounding countryside. Heat islands have been discovered even in villages with as few as three hundred inhabitants and have produced, over a decade, temperature rises of over 0.3°C or more – equal to the global increase for the last hundred years.

The recent availability of a decade of satellite data, of much greater accuracy than data from earth-based stations, also suggests that there may be a heat bias in existing records. Dr Roy Spencer of the NASA space centre in Huntsville, Alabama, has been quoted as saying: "Over the entire ten-year period [i.e. the 1980s] there was no net warming or cooling".[18] Even with ground-station figures, it is possible to isolate periods – 1930 to 70 – when the trend was towards cooling. Those who cast doubt on any warming maintain that the hundred-years record is not long enough to establish a trend, especially given the historical evidence of the Little Ice Ages (1550–1850).[19]

There are also many uncertainties about sea-level rises. The 0.5 m rise, for example, used originally in the WG1 model does not appear great: the size of a small wave. Most of the data come from the tide gauges in sheltered harbours and on jetties. The data average readings, and there is often no compensation for any land movements. There are areas – south-west Britain, for example – where there is geological sinking which raises sea levels, whilst in Scandinavia the land is rising. Some low islands in the Pacific and Indian oceans are certainly currently threatened and others flooded, but by cyclonic surges rather than sea-level rises. Other islands, such as the Shetlands, are rising. Satellite imagery has also contradicted a small number of submarine reports of thinning polar ice.

The observational picture has led to a criticism of global computer modelling which actually relies less on weather observations than on dynamics derived from basic geophysical data, such as the amount of radiation arriving from the sun and the radiation propensities of the atmosphere, working back to temperature, winds and precipitation. In the past, the grids used for GCMs were the size of England, so that local variation was difficult to incorporate. More powerful computers are halving the resolution and also accommodating to "negative" feedbacks such as cloud cover.

The major thrust of the IPCC work has been concerned with the so-called Scenario A – what has come to be known as the "business as usual" scenario. This scenario assumes "few or no steps" taken to limit greenhouse gas emissions, with a doubling of CO_2 by around 2025 and some compliance with international agreements. This scenario is examined in Chapter 3; we shall now examine how mountains fit into all this debate about climate change.

Notes

1. Gribbin, J., *Hothouse Earth* (New York: Bantam, 1989); Gribbin, J. and Kelly, M., *Winds of Change: Living with the Greenhouse Effect* (London: Hodder and Stoughton, 1989).
2. Slade, D. "A survey of informed opinion regarding the nature and reality of a global greenhouse warming", *Climate Change*, vol.16, no.1, (1990), pp.1–5.

3. Intergovernmental Panel on Climate Change, Policymakers' summary of scientific assessment of climate change, Report of Working Group 1 (1990).
4. Strub, R.A., "La Contribution du gaz carbonique au changement climatique", *SEBES*, 1 (1990), pp. 35–40.
5. Lindzen, R., "Generating hot air" *Times Higher Education Supplement*, 30 November 1990.
6. Ibid.
7. Burroughs, B., *New Scientist*, 15 December 1990.
8. Gribbin, J., (1989).
9. Imbrie, J. and Imbrie, K.P., *Ice Ages* (Harvard University Press, 1979).
10. Ephron, L. *The End* (Berkeley, Celestial Art, (1988).
11. Moisseyev, N.N., *The Future of the Environment* – ed. Pitt D.C. (London: Routledge, 1988).
12. United Nations *Climatic and other Global Effects of Nuclear War* (1989); SCOPE, *Environmental Consequences of Nuclear War* (Chichester: Wiley, 1985–6).
13. Pitt, D.C., *Nuclear Free Zones* – ed. Pitt D. and Thompson E. (London: Croom Helm, 1987).
14. *Nature*, 9 November 1990.
15. Lovelock, J., *Gaia* (Oxford: Oxford University Press, 1989).
16. Grinevald, J., "L'Effet de serre de la biosphere: de la révolution thermo-industrielle à l'écologie globale", *SEBES* 1 (1990) pp.9–34.
17. Rens, I. (Personal communication, 1990).
18. *Sunday Times*, 12 August 1990.
19. Grove, J.M., *The Little Ice Age* (London: Methuen, 1988).

Chapter 2

Mountain and Cryosphere Situations

Where do mountains fit into all this? According to the World Climate Research Programme organized by the World Meteorological Organization (WMO) and the International Council of Scientific Unions (ICSU), the physical climate system is made up of four major components:

- the global atmosphere, which is the most rapidly varying component and also the most energetically active, since it is the heat engine that drives the whole climate system and, in particular, the global water cycle;
- the world ocean, which interacts with the overlying atmosphere on periods of months to years within its superficial layers, while the deeper ocean responds over periods of decades to centuries;
- the cryosphere, which comprises the continental ice sheets and ice caps, mountain glaciers and sea ice;
- the land surface of the continents, with its surface and groundwater flow systems that control evaporation from land and water storage in the ground.

Mountains represent a range of systems, even if the popular conception is of ice and snow. At low latitudes in the tropics or the Himalayas, there can be successive zones from luxuriant jungles on the valley floors to arctic conditions on the peaks. Even in the warmest interglacials, high peaks have remained snowcapped. At the height of the ice ages, the ice sheets did not extend into the tropics.

The topography of mountains complicates both temperature and precipitation. Altitude may be associated with temperature decline in isolated peaks, especially in mid-latitudes where there are high wind speeds, but there is a different situation where (as is usual) there are ranges with plateaux, valleys, basins and so on. Nor are linear relationships true of precipitation or snowfall which may reach a maximum below the highest altitude.

The physiography of mountains also has dynamic effects. Mountains may block or deflect the poleward or eastward movement of the weather. As the weather approaches mountains, precipitation may be stimulated as the air rises or collides with the relief. Mountains orientated towards the equator are warmer. Within mountains, deep-cut glacial and river valleys contain microclimates influenced by altitude, slope, aspect, etc. The presence of water in these valleys in lakes, rivers, wetlands and glaciers provides humidity for evaporation and local precipitation, as well as being the source of runoff for the plains below.

In mountains there are extremes of temperature and precipitation. Some of the wettest areas in the world are where monsoonal storms strike, and no longer is this phenomenon to be found only in the tropics. The mountains are a most prominent part of the cryosphere – especially in Antarctica, the most mountainous of continents, where most (98 per cent) of the world's ice is located. Overall, mountains may constitute (according to different criteria) at least 30 per cent of the world's surface, whilst the cryosphere at certain seasons may be double that.

A most significant point is that the mountains are the water tower for the plains below them. The great rivers rise in the mountains, drawing on the immense store of groundwater, much of it frozen. The European mountains, for example, provide about 50 per cent of the water for river systems which include the Rhine, Rhone, Po and Danube (through the Inn). Much precipitation comes from the westerly flow of weather, as well as polar and intertropical convergence zones. Quasi-monsoon conditions are becoming common. In the Alps, this situation may be created by the so-called Genoa effect. The westerlies arriving across the Atlantic are deflected into the Gulf of Genoa. This deflection stimulates a vortex effect which gathers speed and picks up water over the Mediterranean, spinning off eventually back to the Alps as warm, wet, stormy foehn conditions.

The mountains, therefore, are intrinsically dynamic climatic situations changing in space and time, and the effects of any general climate changes are also complex. There have been major trends in recent years, obvious even to the most casual observer. The Alpine glaciers are retreating and have, for example, lost 30 per cent of their overall bulk; some glaciers (Rhine) have diminished by 50 per cent. The duration and depth of snow cover is much less now than a hundred years or even fifty years ago. Then, even the Jura was described as a valley without spring.

More detail is provided by the analysis of glacier and permafrost changes where there are data for the European Alps covering much of the twentieth century.[1] The retreat of Swiss glacier tongues may be correlated with an increase in air temperatures of 0.5°C to 1°C. Admittedly, some small glaciers are advancing, but generally this is because of steep slopes rather than the volume of the glaciers themselves. None the less, there are anomalies where there has been an increase in precipitation – where there is more exposure to wet maritime influxes or impediment to radiation.

The analysis of permafrost and firm snow also shows a complex pattern. Boreholes in the Swiss Alps indicate a possible cooling since 1950, though the dynamics of convective heat transfer and groundwater are not well understood. Some frozen sites in the mid-altitudes appear to be much less affected than higher latitudes, possibly because of different exposures to wind, snow accumulation patterns, melting, etc.[2]

A major problem with interpreting mountain climates is the lack of weather data as well as uncertainties and confusions in what there is. Two major sources are used: historical indicators and observational data. There are differences, however, in results from different methods such as core sediment analysis, tree rings, ice layers, glacial-geomorphological evidence, etc. Sometimes, as in the so-called Little Ice Ages (1550–1850) there is written evidence, though not often before the eighteenth century in the high mountains. In the 1780s Sassure and others established the first observation posts in the Mount Blanc area. Modern observational data may also be scanty even in the most heavily studied areas, since there are still not enough observatories and stations.

An important exception to the absence of data, at least in the Alps, has been the ALPEX experiment, part of a fifteen-year programme between WMO and ICSU.[3] ALPEX, which was initiated in 1982, looked specifically at the so-called Genoa cyclones. The high winds, heavy precipitation and storm surges associated with the Genoa cyclones cause enormous damage. For the project additional observation stations, aeroplanes, ships and satellites were used to gather data. These data revealed how mountains create not only unique climates but unexpected events such as cyclones, and have served as a model for other regions. It should be remembered that with global warming there is more unstable air in the atmosphere and hence a greater tendency to violent storms.

Most predictions for the future in the European mountain region are derived from computer General Circulation Models (GCMs) under doubling CO_2 conditions. These analyses have been applied to European Alpine regions by Ozenda and Borel[4] as well as in the IPCC WG2. Ozenda and Borel start with two computer models, UKMO (United Kingdom Meteorological Office) and GISS (Goddard Institute for Space Studies), but these have too weak a resolution, covering the whole alpine range (8° longitude 5° latitude), missing the variation in the valley regions. Ozenda and Borel use pinpoint data to argue for nuances in the grosser models. They distinguish, for example, the four quarters of the compass. The south will be up to 20 per cent drier with doubling of CO_2, and olives will be grown in the Val D'Aosta as the Mediterranean climate moves up the south-facing valleys. The north-west may be wetter as well as warmer, while continental conditions (e.g. dry, cold winters) influence the eastern boundaries. Altitude will result in upward temperature movements of 150–200 metres per degree centigrade.

Not only temperature and precipitation affect the Alps, but also the quality of the atmosphere. The mountain forests in Europe were long protected from human interference by their isolation, but have not escaped the pollution produced by industrialization and urbanization. Over half the forests of the Alps and the Carpathians are said to be sick and dying prematurely. The most recent figures released at the end of 1990 by the Swiss Forest Institute increased the number of sick trees by approximately 10 per cent. Both acidity and warming have been implicated as reasons for forest dieback. High levels of acidity focus on the Alps. The ability of forest soils to absorb CO_2 or stop erosion is consequently diminished. The warmer, drier climate is also creating a situation where there is more stress on species which are unable to move and adapt to the climatic situation. Ultimately, therefore, major desertification and deforestation may be the result.

Acidity is accentuated if, as is happening in Central Europe, there is a poleward shift of warm air and the mountains find themselves in a zone of tranquil anticyclonic air. The recent problem of the absence of snow in the Alps at the height of the tourist season is largely because the polar front of cold, snow-laden air lies far to the north. Any southward incursions are blocked by persistent high-pressure anticyclones which last for months on end. These anticyclones are literally mountains of air whose resistance is related to the elevated landmasses on which they sit as a giant atmospheric cap. The contrasts with North America, where southward unseasonable incursions of cold air occur, and the European warmth may be partly explained by the geographic orientation of the great ranges on the two continents. The Rockies run north–south, the Alps/Carpathians east–west. There is a major difference between North and South Europe. In general, the North (Scandinavia) is becoming warmer and wetter, whilst the South (Alps, Mediterranean) is drier and colder in winter (Iberia, Atlas).

The global climate shifts affecting mountains are influenced by the ocean. Cold water from the Antarctic may be conveyed to the northern oceans quite rapidly. The oceanic temperature system has been historically – and is assumed to be in the future – stable. The mean ocean surface temperature in the period 1860 to 1880 was essentially the same as it is now, even if there has been warming or cooling in the intervening period.[5] Combining land and oceans, the temperature increase has been only 0.6°C, and only 0.2°C since 1930. Ocean temperatures are influenced by the mixing of water at different depths, and there is a cooling effect of deeper waters. Some futurological models[6] predict a melting of sea ice, which is already very thin in parts of the Arctic and pulsates annually in the Antarctic. There is a fear that if this were to happen, the lower part of the ocean would be insulated from the upper layers. This would allow more rapid heating of the surface; possibly also the destruction of the living organisms which absorb CO_2.

There is also some agreement that increased ocean temperatures may favour the generation of violent storms. It has been suggested that a mean rise in August temperatures of 2.3°C to 4.8°C would increase the intensity of cyclones by 60 per cent though not necessarily their frequency.[7] In some mountain ranges – the Caucasus, for example – very heavy precipitation is the result of these storms. Some currents or upswellings are influences and indicators of climate change. The most famous is El Niño, so-called because fishermen noticed every four years around December that the waters became unusually warm. El Niño refers to the coming of the Christ child with which December is associated. Climatologists noticed that when El Niño appeared, there were very hot and dry summers in the northern hemisphere. The opposite condition was at first called Anti-El-Niño, but Spanish-speakers did not like the anti-Christ analogy so the cold upswelling was called La Niña – little girl. Some researchers claim that either the cold-water conveyor or the upswelling may suddenly flip the world's climate into heat or cold.

The cryosphere contribution to global climate patterns, then, is considerable. Also important is the effect of glaciers on sea levels. Recent satellite pictures have shown the break-up of the West Antarctic Ice Sheet. With warming, huge blocks the size of Belgium have begun to chip off. As they go, the lighter floating ice is more likely to fissure, especially because a huge ice-stream pressure behind the blocks builds up. These ice blocks may set off cold currents but also raise sea levels very significantly. Rises of up to 6 metres have been predicted.[8]

In temperate latitudes at least, alpine glacial melt also raises sea levels – more than Antarctica in some theories, since small ice blocks (as in a cocktail) melt more quickly than large ones. Additionally, polar ice is locked up in colder conditions (see Table 2.1). In some areas (Scandinavia, for example), as the ice has been retreating, the land is rising, but the general effect of sea-level rise is drowning and increased salinization of low-lying areas and groundwaters, especially if there is a greater frequency of storm surges. This is accentuated in many delta regions because the land is sinking. Low-lying countries and islands are most at risk. Great cities such as Amsterdam, London, New York, Sydney, Hamburg, Bangkok and Shanghai may be vulnerable to flooding due to melting ice thousands of kilometres away.

Table 2.1: Sea-level rise causes, 1985-2030 (cm)

Source	Low	Best guess	High
Thermal expansion	4	9–14	18
Alpine glaciers	2	8–12	19
Greenland	1	2–3	4
Antarctica	-2	-2– -3	3
Total	**5**	**17–26**	**44**

Source: Commonwealth Secretariat (1989).

Notes

1. Haeberli, W., "Glacier and permafrost: signals of twentieth century warming", paper presented to the Seattle Symposium on Ice and Climate (1989); Haeberli, W., "Possible effects of climate change on the evolution of alpine permafrost landscape", Ecological Impact of Climate Change Conference, Amsterdam (1989).
2. See note 1.
3. Newson, R., "The ALPEX experiment", *The Courier* (UNESCO), February 1987, pp. 34–6.
4. Ozenda, P. and Borel, L., "Possible impacts of a climate change induced by an increasing greenhouse effect on the alpine ecosystems", European Conference on Landscape Ecological Impact of Climate Change, Lunteren, the Netherlands, 3–7 December 1989; Intergovernmental Panel on a Climate Change: Policymakers' Summary of the Potential Impacts of Climate Change. Report from Working Group 2 (1990).
5. Folland, C., Parker, D. and Kates, F. "Worldwide marine temperature fluctuations 1856–1981", *Nature*, 310 (1984).
6. *Nature*, 29 June 1990.
7. Emmanuel, K.A., "The dependence of hurricane intensity upon climate", *Nature*, 326 (1987), pp. 483–5.
8. Intergovernmental Panel on Climate Change: First Assessment Report (Geneva: WMO, 1990).

Chapter 3

Climate Change and Mountain Regions

In Chapter 1 we set out some of the major perceptions, assessments and explanations of the current situation. We look now at the most probable effects in mountain regions of the most likely scenarios. We basically use the IPCC "business as usual" scenario, nuanced – as Ozenda and Borel have done – for the Alpine region.[1] That is, we assume both a considerable amount of warming and a situation of considerable change and dynamism, in both time and space. Whatever is happening to global climate, there seems to be an overwhelming consensus that there is or will be warming; the debate now is about when and where, as well as the effects.

The concept of "business as usual" is double-edged. First, it is assumed that industrialization and growth will proceed broadly within the parameters of the post-World War II period, even if such a period of tranquillity is rare indeed in history. Secondly, there is a tendency to play down the gloom-and-doom elements of more catastrophic theories. Winners and losers are identified in the different regions, but also it is felt that there is a balancing out of all the actions and reactions, past and present. Certainly in mid-latitude mountain regions there is evidence of more of a steady state than at high latitudes.

Table 3.1 summarizes the regional forecasts in the "business as usual" scenario prepared by IPCC Working Group 1. Even if confidence in these estimates is low, they give some indication of climate changes. The Southern Europe region includes the Alps, Carpathians, Pyrenees, Balkans, Iberian and Apennine mountains, and part of the Caucasus. More precise figures are supplied by Ozenda and Borel (see Table 3.2).

When will all this happen? There is virtually a consensus that CO_2 will double within the next century, and the different speeds of this process are the starting point for any models. The future

Table 3.1: Regional changes: "business as usual" estimates by Working Group 1

Changes from pre-industrial, i.e. *c*.1750
The estimates are based on high-resolution models, scaled to give a global mean warming of 1.8°C consistent with the best estimate (2.5°C) of climate response to greenhouse gases. With the low estimate value of 1.5°C, these values should be reduced by 30%; with a high estimate of 4.5°C, they should be increased by 50%. Confidence in these estimates is low.

Central North America (35°–50° N 85°–105° W)
The warming varies from 2° to 4°C in winter to 2° to 3°C in summer. Precipitation increases range from 0% to 15% in winter, whereas there are decreases to 5% to 10% in summer. Soil moisture decreases in summer by 15% to 20%.

Southern Asia (5°–30° N 70°–105° E)
The warming varies from 1° to 2°C throughout the year. Precipitation changes little in winter and generally increases throughout the region by 5% to 15% in summer. Summer soil moisture increases by 5% to 10%.

Sahel (10°–20° N 20° W–40° E)
The warming ranges from 1° to 3°C. Area mean precipitation increases and area mean soil moisture decreases marginally in summer. However, there are areas of both increase and decrease in both parameters throughout the region, which differ from model to model.

Southern Europe (30°–50° N 10° W–45° E)
The warming is about 2°C in winter and varies from 2° to 3° in summer. There is some indication of increased precipitation in winter, but summer precipitation decreases by 5% to 15%, and summer soil moisture by 15% to 25%.

Australia (12°–45° S 110°–155° E)
The warming ranges from 1° to 2° in summer and is about 2°C in winter. Summer precipitation increases by around 10%, but the models do not produce consistent estimates of the changes in soil moisture. The area averages hide large variations at the subcontinental level.

Source: IPCC, Policymakers summary of scientific assessment of climate change. Report of Working Group 1 (IPCC, 1990).

scenarios we describe are based on this doubling of CO_2 and the presumption that when it occurs there will be global warming of at least 1°C – i.e. an escalating upward curve – even in the most conservative estimates. The possible climate changes are summarized in Table 3.3.

Table 3.2: Seasonal climatic changes in the Alps range according to the GISS model

	Temperatures (8°C)		Precipitation (mm/day)	
	Western Alps	Eastern Alps	Western Alps	Eastern Alps
Winter	+4.5	+4.3	+0.45	+0.45
Spring	+4.2	+4.1	+0.5	+0.65
Summer	+3	+3.8	+0.3	+0.3
Autumn	+4	+3.8	+0.3	+0.3
Total	+15.7	+15.5	+15.5	17
Mean	+3.9	+3.9	+3.9	+4.2

Source: Ozenda and Borel (1989)

The World Meteorological Organization and United Nations Environment Programme have established three basic scenarios: a fast track – envisaging a doubling of CO_2 concentration from 300 parts per million (ppm) to 600 ppm by the year 2030, a medium track – by 2060; and a slow track – by 2090. The other greenhouse gases have similar projections. Methane and nitrous oxide are linked to agriculture and the assumption that the human population will double in the next hundred years, CFC increases are rather to be explained by consumer demand for specialized products, such as refrigerators. The fast-track scenario is explained by the following causal factors. First, it is assumed that developing countries will expand their use of fossil fuels at a rapid rate. Between 1973 and 1980, developing-country energy consumption grew at a rate of 62 per cent per annum compared to 0.5 per cent in the developed world. Poorer countries have less access to energy sources which are less polluting, and it is assumed that there is a lesser effort on energy-conservation measures.

An added factor is the greater increase in population numbers

Table 3.3: Possible climate changes from doubling of CO_2

Large Stratospheric Cooling (virtually certain). Reduced ozone concentrations in the upper stratosphere will lead to reduced absorption of solar ultraviolet radiation and, therefore, less heating. Increases in the stratospheric concentration of carbon dioxide and other radiatively active trace gases will increase the radiation of heat from the stratosphere. The combination of decreased heating and increased cooling will lead to a major lowering of temperatures in the upper stratosphere.

Global Mean Surface Warming (very probable). For a doubling of atmospheric carbon dioxide (or its radiative equivalent from all the greenhouse gases) the long-term global mean surface warming is expected to be in the range of 1.5 to 4.5°C. The most significant uncertainty arises from the effects of clouds. Of course, the actual rate of warming over the next century will be governed by the growth rate of greenhouse gases, natural fluctuations in the climate system, and the detailed response of the slowly responding parts of the climate system, i.e. oceans and glacial ice.

Global Mean Precipitation Increase (very probable). Increased heating of the surface will lead to increased evaporation and, therefore, to greater global mean pecipitation. Despite this increase in global average precipitation, some individual regions might well experience decreases in rainfall.

Reduction of Sea Ice (very probable). As the climate warms, total sea ice is expected to be reduced.

Polar Winter Surface Warming (very probable). As the sea-ice boundary is shifted poleward, the models predict a dramatically enhanced surface warming in winter polar regions. The greater fraction of open water and thinner sea ice will probably lead to warming of the polar surface air by as much as three times the global mean warming.

Summer Continental Dryness/Warming (likely in the long term). Several studies have predicted a marked long-term drying of the soil moisture over some mid-latitude interior continental regions during summer. This dryness is mainly caused by an earlier termination of snowmelt and rainy periods, and an earlier onset of the spring-to-summer reduction of soil wetness. Of course, these simulations of long term equilibrium conditions may not offer a reliable guide to trends over the next few decades of changing atmospheric composition and changing climate.

High-Altitude Precipitation Increase (probable). As the climate warms, the increased poleward penetration of warm, moist air should increase the average annual precipitation in high latitudes.

Rise in Global Mean Sea Level (probable). A rise in mean sea level is generally expected due to thermal expansion of seawater in the warmer future climate. Far less certain is the contribution due to melting or calving of land ice.

Source: Leggett, J. (Ed.) *Global Warming – The Greenpeace Report* (Oxford: Oxford University Press, 1990).

in developing countries. Infant mortality rates are falling (below 50 per thousand), life expectancy is rising (now over sixty years) and consumer demand (especially for cars) is expanding along with increased *per capita* GNP and expectations of Western life styles. The increasing movement of populations to the cities and industrial concentrations creates pollution black spots. A new area of energy increase has emerged recently with perestroika in Eastern European countries. There is a very strong demand for Western consumer products, especially motor vehicles. At present, the car pool in Europe is increasing by 5 per cent annually, but this rate may double if Western car manufacturers are successful in their marketing campaigns to provide every East European family with a car. The medium and slower tracks assume some kind of restrictions in either/or population growth and consumer demand increase or increased effectiveness of energy-conservation measures. Historical experience certainly indicates that there may well be "natural" checks to population increases – wars, the AIDS pandemic, etc. – as well as a growing success for family-planning programmes.

Global warming itself will save energy to some extent, since there will be lower heating bills. Improved energy-conservation technology (house insulation, vehicle catalysers, solar-energy cars) will all help, especially if any propaganda is stiffened with legislation. The green ethic may well prompt people to look for ecological solutions for their daily energy needs – walking, bicycles, wearing sweaters, reducing consumption of fast foods, using less wrapping. CFCs may well be outlawed completely, since they do not occur naturally. There must, however, be much uncertainty about the degree to which governments will move to control emissions or climate change. The Ministerial Declaration issued after the Second World Climate Conference in November 1990 (Appendix II), was quite vague about commitments to specific levels, despite NGO pressure. Some agreement may, however, emerge on stabilization at 1990 levels by the year 2000 and a 20 per cent reduction by 2005 (see Chapter 6).

A most important point is that even if there were stabilization at present levels, there would still be increases in key gases which have long lives (CO_2: 200 years; CFCs: 100 years; see Figure 3.1).

Estimates of the effects of warming on mountain climates are, of course, as speculative as those for other regions. It is presumed that with warming there will be less permanent snow, increased glacial melt, and a higher tree line. But in some models there is an increase in precipitation too – for example, in the wetter inner axis of the Alps. There will also be major effects on the permafrost, whether in the mountains or the tundra zones. In the mountains, the melting of frozen materials in the upper layers increases the risk of severe rock falls from the debris. Avalanches will generally increase as the warmer weather creates spring snow conditions.

Figure 3.1: Greenhouse gases and scenario effects

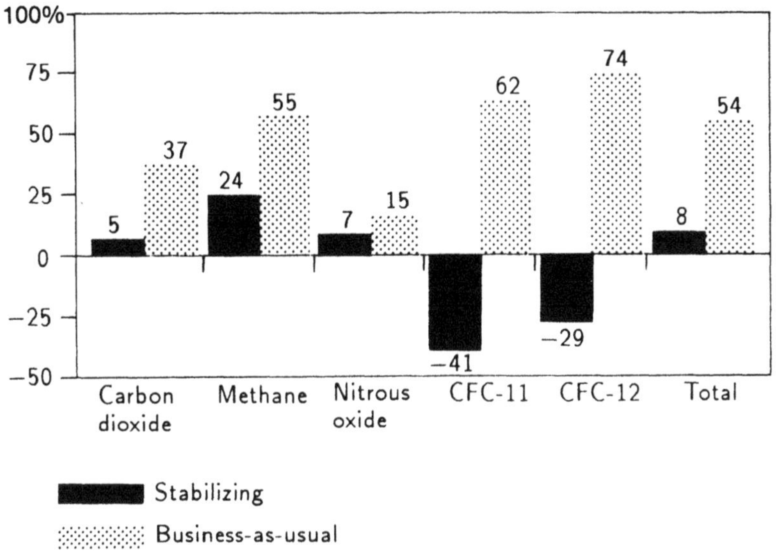

Source: Leggett, J. (Ed.) *Global Warming – The Greenpeace Report* (Oxford: Oxford University Press, 1990).

There are some counter-intuitive trends. The key factor may be less temperature than precipitation, and predictions for parts of the European region are for wetter conditions and more soil humidity. For example, some mountain ranges (e.g. the Caucasus) are already experiencing more, rather than less, severe winters. Very heavy snowfalls (up to 10 metres have been reported) in the Caucasus are associated with monsoonal conditions in the Black Sea which falls as snow because of the high altitudes of the range.[2] By contrast, the Alps are experiencing less winter snow because of the stability of dry, if cold, anticyclonic conditions in winter.

Whilst the melting of the high-latitude permafrost will enhance the greenhouse effect through the release of methane gases, other melts may decrease it by providing more evaporation and hence more clouds. Most glaciers are melting and retreating, as they have since the last ice age, but a number in the Alps are starting to advance, and advances may be triggered by warming.

A major effect of warming is on the water system (see Table 3.4). Runoff, for example, is exponentially sensitive to temperature rises when accompanied by aridity. A 1°C to 2°C rise in temperature accompanied by a 10 per cent fall in precipitation could produce a 40–70 per cent reduction in runoff. Conversely, in regions where there is increased precipitation (North European mountains) there may be excess water or ice and consequent flooding dangers, as well as extra pressures on water-supply systems and hydro dams. Erosion, especially if there is massive deforestration, will result in increased amounts of silt being washed down into the river system, sometimes causing huge floods far downstream as in the Himalaya–Indo–Gangetic system. In any case, changes in snow accumulation and melt patterns will result in the redistribution of major bodies of water, changes in the regimen of streams and rivers and in the moisture status of soils.[3]

Sea-level rises are not usually directly related to mountains, though lake and river levels are. Hydrological variation may greatly affect energy and water supply, navigation, health and conservation. In general, glacial melt may initially increase river and lake levels, but in the long term drying out will be more common. With a warming regime, there may be more hydrological disasters. Ice- or moraine-dammed lakes may empty, as happened near Salzburg in 1932 when a plug shifted. Wetlands may drain when the ice floor melts. A general melting of the permafrost, wherever it occurs, will lead to difficulties for transport and housing, as well as facilitating slips and avalanches.

Several climatic effects on vegetation have been picked out in global climate modelling scenarios, with many effects on mountains. First, the relatively greater warming predicted in higher latitudes will allow a northward advance of forests by about 150 kilometres per degree of temperature rise. A similar effect to this latitude increase has been noticed in altitude too. In the European Alps, a 1°C rise results in the tree/crop/vine line rising by about 150 metres. Thus, if temperatures rose by 3°C, the Alps would expect a climate and vegetative regime similar to the Pyrenees

Table 3.4: Climate change and water resources

Climatic impact water function	Annual runoff (increase or decrease)	Runoff frequency (increase or decrease)	Runoff seasonality (less even)	Sediment production (greater)	Sea level (rising)
Water supply reservoirs	Increased yield in humid areas Reduced yield in semi-arid areas	Increased yield more reliable in humid areas Reduced yield and reliability due to more drought in semi-arid areas	Less even distribution, so reduced yield and reliability generally	Increased sediment, so loss of storage and reduced yield	Saline intrusion of aquifers Flooding of reservoirs
Hydropower reservoirs	Increased generation in humid areas Reduced generations in semi-arid areas	Increased generation and firm load in humid areas; reduced firm load in semi-arid areas	Reduced generation and firm load	Loss of storage and reduced generation from sediment	
Flood protection	Increased protection required in humid areas Protection in semi-arid areas; need to cope with infrequent severe floods	Greater flood frequency in humid areas, increasing flood hazard Fewer wet years in semi-arid areas but flood severity could be greater	More frequent floods in humid areas; greater wet-season runoff in semi-arid areas; flood hazard	Loss of reservoir storage; increased deposition; increases of flood hazard	Serious hazard in low- lying areas; sea flooding frequency and severity increased
Environment resources and water quality	Quality improved in humid areas Quality reduced in semi-arid areas	Quality improved in humid areas Quality impaired in semi-arid areas	Dry-season water quality affected in semi-arid areas	Water quality impaired by sedimentation	Groundwater contamination by salt water Saline infiltration into sewerage system
Navigation	Improved river and lake navigation in humid areas	Reduced risk of drying up in humid areas		Sedimentation reduces navigation	Affects viability of low-lying port installations

Source: Commonwealth Secretariat (1989).

today, whilst the Scandinavian mountains might resemble the Alps, especially as the warming increases with latitude.

A second effect is a northward shift in the precipitation belts associated with the Intertropical Convergence Zone and the Polar Front. The effects in European mountain zones may be further to increase precipitation in northern zones, particularly Scandinavia, and hence further to enhance the potential for forest and agriculture expansion. By contrast, in the South – e.g. the Alps and Carpathians – there are forecasts of decreased precipitation, especially on southern slopes. There will be increased dryness, and the growth and expansion of existing forests and cultivation will be restricted. Precipitation patterns may also be influenced by monsoonal tendencies, with very heavy downpours on exposed slopes leading to erosion and flooding.

A third major effect will be changes in patterns of soil-water availability. With temperature increases, the level of evapotranspiration also rises. Any soil-water deficiencies are most likely to be felt in the warmer, drier regions of the Mediterranean mountains, Carpathians and south-facing slopes of the Alps.

Most discussions of climate change are concerned with the atmosphere and warmth, but it is important to recognize the changing climate of other ecosystems, particularly soils. Agriculture is dependent on this pedoclimate. The map of changing soil climates does not necessarily mirror atmospheric changes. Soil type, texture, structure, position, depth are also significant. Soil humidity is a key factor, since optimum conditions require a balance between the extremes of desiccation and waterlogging. Although the increased dampness in some areas may be a positive factor, levels of acidification may remove any advantage. Mountain soils are not well adapted to pedoclimate changes. Fragile, unstable and stony soils cannot easily withstand major changes in temperature and precipitation regimes.

Both warmer and drier conditions reduce the amount of snow. A one degree rise in average annual temperatures, combined with winter drought, may reduce the duration of snow cover by 50 per cent at 1,500 metres in the Alps, with enormous consequences for the tourist industry. The present vegetation may be placed under stress by these climate changes – not least because pests or disease may survive more easily in the milder conditions. There is also a greater risk of fire and increased mortality, especially in the overmature stands which are common in the Central European mountains.

If there is warming, one of the first changes will be a shift in the situation of the ecological belts. Each belt is often defined by characteristic flora but additionally has its own soils, fauna, land use and microclimate. The whole can be regarded as an interrelated system defined within an altitudinal belt, as shown in Figure 3.2. These belts have an approximate relief of 700 metres. Thus belts would roughly move up to succeed each other and the northern Alps would take on the appearance of the southern (European) Alps. In fact, most flora do not have such an adaptive range, so that even if over the long term there was a succession in the short term there is more likely to be degradation and disappearance of species. The problem is particularly acute for trees, which cannot migrate as quickly as herbaceous species and fauna.

Combining temperature with precipitation may mitigate some of the apparent effects. Ozenda and Borel consider that some species in the Pre-Alps area would rejoin their niche because by moving up in temperature they would receive more precipitation. In the inner alpine axis the same authors consider that the 10 per cent increase GISS forecast would have little effect, but the decrease forecast would be devastating, creating steppe conditions, a desiccation enhanced by increased temperatures and evapotranspiration.

Atmosphere changes involve not only new patterns of temperature and precipitation but also the direct effects of the greenhouse gases themselves. CO_2 increases may have a fertilizing effect, but the jury is still out on the evidence, at least as far as forests are concerned.[4] There is some indication that the carbon fixation biochemistry of trees – the so-called C3 group – which predominates in cool-temperature zones, both latitudinally and altitudinally, may be favoured. The specific effects of different scenarios on the forests will be discussed in the next chapter.

Temperature rises and CO_2 concentrations may favour forests, especially in the cooler zones, but the same cannot be said for other greenhouse gases and pollution in general. Studies of Alpine forests have shown that over half are affected by acid pollution and more are exposed to risks of pests and disease, as well as having a lessened life expectancy, though the interrelationships have been difficult to prove. The major causes of forest decline have been described as in order of rank, ozone, acid deposition, other gaseous pollutants (SO_2, NO_x), excess nitrogen deposition, and growth-altering organic chemicals.[5] Lower-altitude ozone (not to

Figure 3.2: Altitude zone schema (Alps)

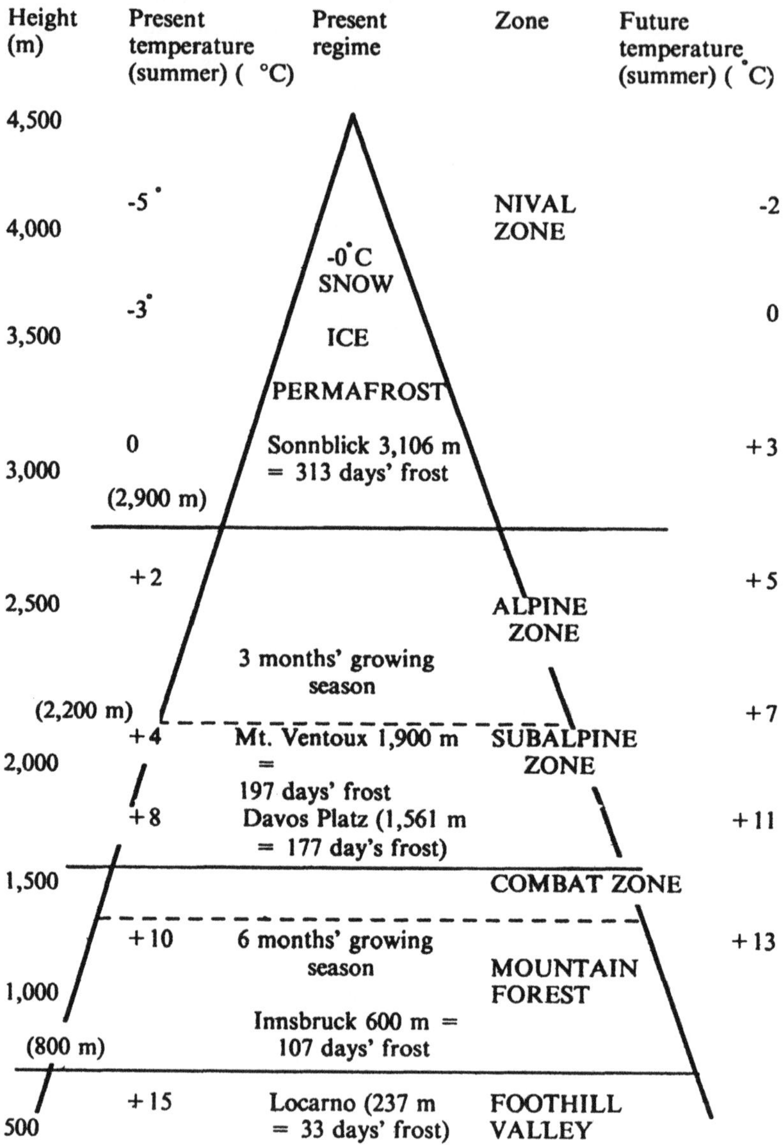

Source: Alp Action archives.

be confused with high-altitude natural ozone) can lower the rate of photosynthesis and directly damage trees – conifer needle surfaces, for example – weakening resistance against disease and speeding up drying out, but also assisting in the formation of acid pollution. Ozone has been cited as a prime cause of forest decline in the Alps where levels of SO_2 have been stable or even declined in the recent years, though low, continuous doses have been implicated.

The decline of the forests constitutes a vicious cycle. Pollution and warming temperatures were part of the reasons why the trees were dying. The forests were, therefore, less able to resist diseases, pests and severe climate. For example, the huge storm in the abnormally warm and turbulent conditions of February 1990 wreaked havoc in the Alpine forests. In the cantons of Glarus, Grisons and the Unterwald, the windfalls were more than 2.5 per cent of the woodstock – three times the annual exploitation, though this was possibly an exceptionally violent tempest. Winds of 269 km/h were recorded at Grand St Bernard. The abnormally warm conditions and weakened trees favoured pests, notably the bostryches, which were estimated to destroy 285,000 cubic metres (1985–6) of spruce, or 10 per cent of the number of conifers exploited. The maps of windfall incidence and bostryche infestation show considerable similarities.

The most recent figures on damage in the forests indicate a worsening situation, especially in the Swiss Alps.[6] There is a rough-and-ready scale by which damage is measured according to the amount of defoliation. Between 1985 and 1990 the number of trees in Classes 1–4 (11–100 per cent defoliation) rose in Swiss mountain regions from 42 to 68 per cent of all trees, whilst Classes 2–4 (26–100 per cent defoliation) nearly doubled (11–21 per cent). The increases in affected trees was not a smooth curve; 1988 and 1989 recorded only 51 per cent and 49 per cent damage (Classes 1–4). On the plains, the number of affected trees (Classes 1–4) rose from 27 to 48 per cent (1985–90). The higher Alps region had more affected trees in Classes 1–4 (72 per cent) than lower mountain regions, such as Jura (65 per cent, Southslopes (68 per cent) and Pre-Alps (57 per cent). The high figures for the mountains and the association with altitude are partly to be explained by the increasing predominance of conifers with height. Conifers have had a higher rate of damage than deciduous trees: 63 per cent compared to 55 per cent (Classes 1–4).

The complexity of interrelationships must be emphasized. A

recent Swiss study has placed more emphasis on climate change than on pollution, whilst recognizing than any interpretation must be based on extensive research (much not available) of soil, vegetative and other physiological processes. In this study there seemed to be some relationship between warm and dry conditions in the summers before the deterioration was detected.[7]

Ozone depletion is another effect of pollution with particular effects in mountain areas. Chlorofluorocarbons are a family of substances widely used as refrigerants, solvents, foam-expanding agents and aerosol propellants. When released into the atmosphere, they slowly diffuse to the upper stratosphere, where their molecules are broken down by radiation, producing chlorine which strips an atom from ozone, making it normal oxygen. Ozone "holes" have been discovered over colder mountainous areas like Antarctica, but some thinning has also been found over the Alps. Ozone deficiency has a major effect because normally the ozone layer excludes ultraviolet radiation, which can damage living tissues. Skin cancers are more common in mountain areas and there is also an effect on sensitive crops.

The detailed analysis of the effects on forests follows in the next chapter, but it is important to note how different climate-change scenarios affect the incidence and distribution of pollution. Acidification maps of Europe show a high incidence over Alpine areas. This high incidence, attaining critical proportions, persists in future scenarios even after emission reductions[8] September 1990). Higher temperatures may lead to more rapid chemical processes. Water-vapour distribution is important, since it is necessary for both acidification and deposition. Prevailing wind directions and forces determine where depositions will occur. Mountain topography is important too. There may be higher deposition rates on windward slopes, in wet valleys or in those areas susceptible to acid mists. Probably because of cloud and mist formation in certain altitude bands (800–1200 metres in the Alps), there is also a concentration of acid deposition and tree dieback.

Of considerable importance is the scale and speed of climate change. The warming that is predicted even in the most conservative estimates to the year 2030 would make the earth hotter than at any time in the last 120,000 years. Nor are these speeds uniformly distributed over the earths surface. Such changes have been accommodated much more slowly in the past, and existing ecosystems cannot easily cope. The present rate of change is estimated to vary between ten and a hundred times that previously encountered. For example, trees with light seeds may disperse only 2 kilometres per annum. It is not certain what will happen in the ecozones between broadleaved boreal forests and tundras to the north or conifers to the south, or in the altitudinal analogues. In those areas where precipitation is increased, wetlands may enlarge. Conservation experts fear that many species will not be able to adjust to such abrupt changes and may become threatened.

On the other hand, most studies have been carried out latitudinally rather than altitudinally. Vertical dissemination is much easier than horizontal, and the common mountain winds may help. Trees with winged seeds (maples, ashes) can spread over several hundred metres' altitude in a season. Modern silvicultural techniques may also assist. Those species whose spread is vegetative are in a different category, and in any case competition will be accentuated.

Key economic sectors of mountain activities will also be affected by the "business as usual" scenario. First, in mountain agriculture the number of growing days can be expected to increase at altitude, so improving the food situation, especially for those mountain communities dependent on subsistence – an effect possibly enhanced by CO_2 stimulation. Although poor mountain communities are mainly to be found in the Third World, some poor communities in the European mountains would also benefit – in Eastern Europe, for example. New pests may be introduced with warming, especially where winter freezing does not reduce their numbers. As precipitation is likely to come in more frequent bursts, erosion and avalanche risks will be increased. Not all regions, of course, are predicted to have precipitation increases. For example, the Mediterranean mountains and adret slopes of the Alps may well experience summer drought, and yields have been predicted to fall by as much as 15 per cent.

The whole pattern of crops and fruits may change in reaction to small but critical climate changes. For example, many fruit trees depend on winter chilling to initiate or accelerate the flowering process. The range of other crops may be restrained by the number of frost-free days. Equally important are the number of days at wilting point – when all the available water capacity in the soil for plant growth is depleted. Wind and light conditions require adaptation. All this may necessitate a complete relocation of horticultural and arboricultural practices. There will be a major effect on conservation.

First, in the mountain zones of Europe there are a number of altitude zones, each with its own specifically adapted and often threatened fauna and flora (see Figure 3.2).

On the summits is the nival zone, basically a zone of permanent snows. In the Alps, this zone extends at present from 2,500 to 4,000 metres altitude. Contrary to the popular view, this zone is rich in wildlife; important species are lichens, glacial ranunclus, the heresian primrose, the brown and orange mountain butterfly, the snow finch, the royal eagle, the glacial flea – to name only a few. Many nival species are already threatened. If the snow and freezing limit were to shift upward by 700–1,000 metres (i.e. +3°C temperature rise) the nival zone would practically disappear, being confined in summer to small regions on the Mont Blanc massif. The median height of the Alps summits is around 2,000 metres, barely enough to attract winter snow for more than a month or two, let alone year-round cover. Certainly some species can migrate (notably birds), but the pressure of population concentrations will lead to increased species decline and further loss.

Below the present nival level, between 1,800 and 2,800 metres, is the Alpine zone proper. This zone, usually defined as having a growing season of fewer than 90 days, is characterized by small plants and grasses that can grow rapidly in the short season. The stunted herbaceous willow, the yellow saxifrage and the edelweiss are typical, but there is an abundance of fauna – the alpine ptarmigan, the black salamander, the Apollo butterfly, the viviparous lizard, the snow campagnol, the marmotte, the Siberian cricket, etc. Many of these species have already suffered from human incursions into the alpine zone – not only high altitude ski resorts but also pollution. Perhaps because this region has frequent mists and cloud, there are high deposition rates of toxic substances. Some plants concentrate toxins. The Chernobyl effects were very obvious, and still persist in this zone. Some of the fauna – the black salamander, for example – are as sensitive an indicator of dangerous thresholds as canaries used to be down the mines.

Like the nival zone, the alpine area will become compressed with climate change. A 700–1,000 metre rise in the limit would eliminate the zone in many mountains, especially as the "combat" and forest zones below push upward. Some species may be able to migrate, but adaptation is difficult because these zones are distinguished not only vertically but also horizontally by the dissection of the mountains into valleys. Given the different precipitation patterns, the alpine zone will be at different levels on the adret and ubac slopes. The wetter ubac slopes favour forests to higher levels (Figure 3.3). As precipitation levels are likely to increase on the northern slopes in the Alps, there is likely to be more encroachment on the alpine zone.

Figure 3.3: Aspect and vegetation

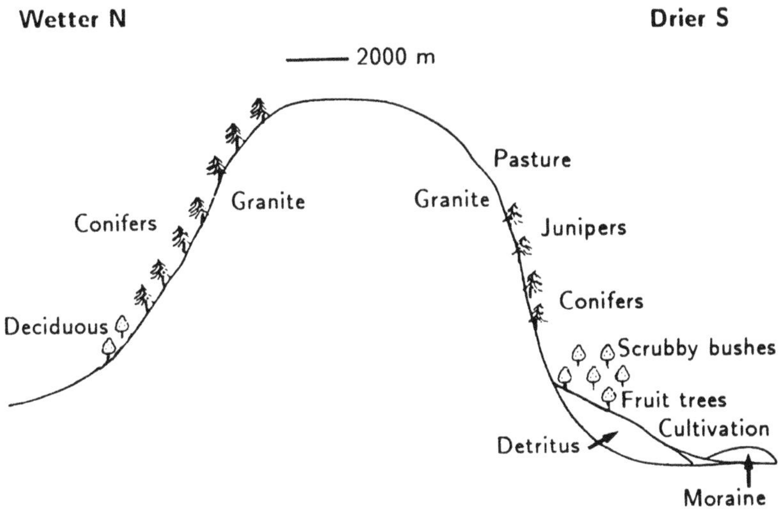

According to Ozenda and Borel, the higher ranges of temperature increase estimates (e.g. 3.8°C) would virtually eliminate the alpine and nival zones from the Central and Southern European ranges. Some remnants may remain in the Pyrenees, and areas in the Alps would be greatly reduced. The alpine zone might disappear completely from the Western Alps (south of Pelveux), the Dolomites, the Eastern Alps and east of the Hohe Tauern. Two zones only would remain: the Mount Blanc massif and the East Tyrol – at best 25 per cent of the existing nival and Alpine zones. With this disappearance of the alpine and nival zones, at least 150 vegetative species would be threatened or disappear. The fauna would be less affected in terms of immediate mobility, and already there are studies of ibex or migrating birds adapting.

But this fauna is, of course, dependent on the vegetation, which would either disappear or for which there would be increased competition.

If the temperature rises dramatically and precipitation decreases, a progression of Mediterranean ecosystems may be expected. Such an ecosystem is characterized by a shrubby-type vegetation (maquis). The temperature would have a high annual mean (+13°C), absence of winter frosts, high aridity (especially in summer), and the limit for olive cultivation may reach Aosta.

Below the alpine zone are the regions where the forests begin, often divided into two subzones: the sub-alpine, at present 1,200 to 2,000 metres and the mountain or forest zone 600 to 1,200 metres. Conifers dominate in this zone; also to be found are deer, famous birds such as the tetras, rhododendrons, the lynx, and snakes. In many parts of the Alps, meadows are heavily utilized for transhumant dairy agriculture as well as for ski lifts.

It is presumed that the upper-altitude levels of the forest will increase by as much as 1,000 metres. If the nival and alpine zones are losers in conservation terms, the forest zone is in one sense a winner. There are, however, some cautionary notes to be sounded. There is a lag in natural regenerative response to temperature rises. If, as seems likely in much of the Alps, there is a decline in precipitation as well as a rise in evapotranspiration, especially in the growing season, there will be a drying out of forests and a greater susceptibility to disease. Historical studies derived from tree-ring analysis have shown that the present is a time of a lesser tree-growth cycle. Forest dieback may be accentuated. Given the expectation that it will be difficult to control pollution emissions, the forests may decline unless there are vigorous programmes of replanting with the most resistant species.

Below the forest zone are the valleys, usually marked by the limits of the vine on adret slopes which in former times rose to 1,200 metres in sheltered parts of the Valais. These abandoned vineyards may again be viable, but the limits and types of cultivation will be determined by social, economic and political factors rather than climatic. The political future for many countries involves internationalization and rationalization of agriculture and other industries, with a reduction of Government subsidies. This would not favour any expansion of mountain activities, which are in many respects marginal and peripheral.

There may, however, be counter-trends, at least in some regions.

Countries such as Switzerland have been reluctant to enter into international agreements and may favour national autarchy. Here there is strong pressure from peasant groups to continue subsidies. This policy could favour mountain farmers, who would have to supply most of the national market with dairy products, cereals, fruits, wines, etc. What is more, Europe's "green wave" may lead more people to decide to go and live in the mountains and may also mean an increased demand for "organic" mountain products.

Many mountain communities are above the "fog" line: in the sun in winter when the plains cities have cold anticyclonic smog. The smog factor is increasing as more cars come into the cities. By contrast, more mountain towns restrict traffic on the model of Zermatt. Telecommuting may lead to a decentralization of work. Lighter service industries are possible in the mountains, and already centres such as Davos are attracting many businesses. The "grey" revolution whereby the majority of the population are ageing will also favour an increase in the use of mountains for retirement and health care, especially in the sunny, warm mid-altitudes. Traditional health stations such as Leysin may experience a renaissance.

High-altitude stations, although they have less snow, may capitalize on the search for the sun amongst tourists and residents, especially as tropical and marine tourism may decline with growing pollution, health hazards and political instabilities. The advent of cultural tourism, with musical events and artistic exhibitions and the rediscovery of Alpine history, may favour a growth rather than a decline in visitors. Green ideas will stimulate walking and nature studies. Soft tourist activities like cross-country skiing may also expand, with a heavy concentration on the permanent snows of the glaciers.

Traditional "snow-related" activities are bound to decline. The industry depends on an adequate snow cover for four or more months. For Alpine skiing a minimum of a metre is desirable. Small meterological changes may make all the difference. If December is warm, there may not be a snow base for any later precipitation to settle on. In February it is often cold but dry. In addition to tourism, a number of specific socioeconomic elements with many knock-on effects have been identified in the reports of IPCC Working Group 2 (1990). This group concluded that "projected changes in climate will dramatically reduce the areal extent and volume of the terrestrial cryosphere". In general this Working Group plumped for a higher range of temperature increases than Working Group 1.

First, the global climate models indicate a major decrease in mid-latitudes of the extent and volume of snow cover. This results in a reduction in the albedo effect and therefore enhanced warming. The loss of snow cover threatens plants (including valuable crops) sheltering under the snow, and hibernating animals. There would probably be much less water available for drinking and irrigation as well as for hydropower and waste management. In Europe, the Alps and the Caucasus have been singled out as especially sensitive – as have the plains which depend on mountain melt. The discharge in a river such as the Rhine has a major origin (about 50 per cent) in the Alpine headwaters. Transport is also affected where, for example, snow roads may become impassable when there is only mud.

Ice sheets and glaciers are already melting in mid-latitude, though the increased precipitation in parts of Eurasia (East of the Urals) may lead to an equilibrium. In the last hundred years, the Alpine glaciers have lost about a third of their extent. The prediction in Austria is that if there is a 3°C rise in temperature by 2050, there will be an additional 50 per cent reduction in glacier size. The Soviet Arctic glaciers may disappear in 150–250 years. The increased flow from glaciers will help hydro-electricity generation, or at least offset any decline in precipitation at first, but shortage of meltwater will become a severe problem later.

A more dangerous consequence of melting may be the instability created in the debris zone on the surface of mountains, leading to catastrophic rock slides as well as avalanches burying roads, houses and disrupting water supplies. In addition, very heavy rains and snows associated with climate change lead to floods and avalanches. The poleward movement of the tropical convergence zone and monsoon conditions increases the likelihood of severe storms, erosion and flooding, which has already been seen in the Alps and Caucasus. Permafrost regions in the northern mountains, especially in valley bottoms and intramontane basins, may be especially vulnerable to melting, releasing methane and to a lesser extent CO_2 (further enhancing the greenhouse effect), eroding the soil, creating ponding (melting of top layer) and landslides – all affecting the integrity of transport and building systems. The effect of global warming may be very dramatic in the permafrost zone. According to Soviet studies, the southern boundary of the

Russian permafrost may move 500–700 kilometres north with a 2°C temperature rise.

Catastrophic events may be linked with climate change, especially when changes are sudden. In the Middle Ages the cold conditions of the thirteenth and fourteenth centuries led to poor harvests, malnutrition, lowered immunity and less resistance to devastating plague. Very heavy rainfalls and snowfalls in our time may put extra pressure on engineering facilities, particularly dams. When these dams have been weakened by earthquakes or "concrete cancer" (the deterioration of building materials) a catastrophic situation may exist. Earthquakes may be a particularly dangerous triggering factor, leading to breaches in dam walls or leaks from nuclear power plants. The European mountains are part of the Alpine–Himalayas seismic belt, which is still active. Even Geneva is classified as a medium-risk zone. In the fourteenth century, the Alpine region saw one of the ten greatest known earthquakes when all the houses and churches fell down. A recent newspaper article presented the following scenario:

> It is two minutes to seven on the evening of Saturday 12 November in the year ? An earthquake of 7.1 on the Richter scale strikes the Valley of Hérens for 11 seconds. The violence of the shock ruptures the Grande Dixence, the largest dam in Switzerland, letting loose 400 million cubic metres of water. At this period of the year the accumulation lake is at its maximum level. The town of Sion, only 20 kms away as the crow flies, is wiped from the map in a gigantic tidal wave.[9]

The health effects of pollution and climate change are also significant and have been documented by the World Health Organization. In a celebrated incident in 1990, WHO presented to a UN meeting a confidential document concerned with the long-range transport of pollution which was leaked to the press by the NGO Greenpeace.[10] The main effects concerned a lessened life expectancy (ten years) if pollution levels were to continue and increased morbidity, especially cancers, pulmonary and cardiovascular ailments. There are data showing how increased temperature greatly raises risks of strokes. Some of these effects are accentuated at altitude. For example, ozone concentrations are higher at altitude and may reach dangerous proportions in the summer, inducing cancers. In winter toxic fogs may settle in valley bottoms, worsening heart and respiratory conditions. Heavy metals may deposit in the soil, entering the food chain (see Figure 3.4).

Figure 3.4: Health effects of climate change

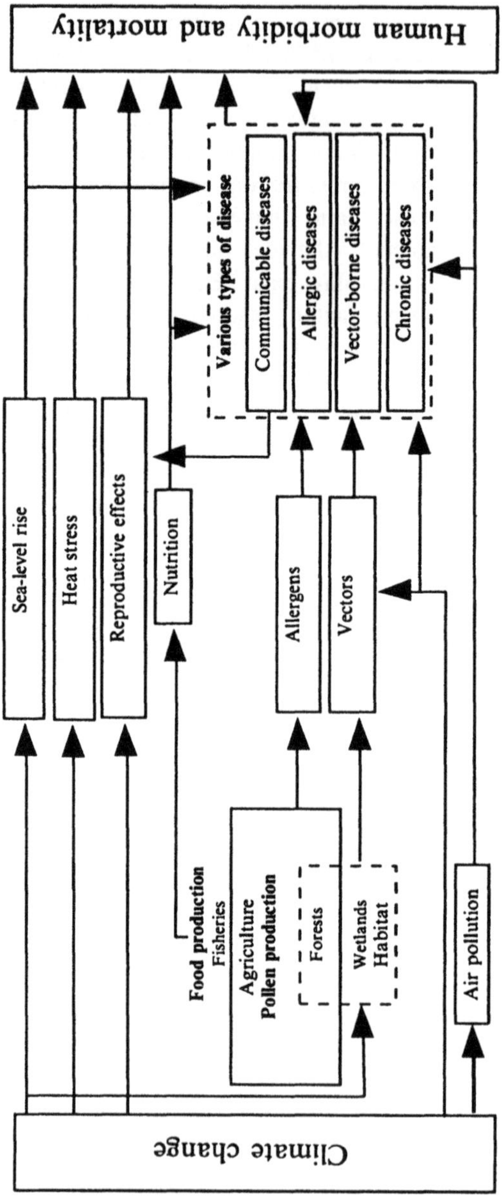

Source: World Health Organization

The degree of effect is also related to the country or region's socioeconomic status. As a general rule, poorer countries or regions are less able to withstand disadvantageous climate change. It is possible to construct a matrix (see Figure 3.5) plotting mountains by their socioeconomic status and possible climate-change scenario. This matrix, however, should be taken as a very rough guide only. Many mountain regions as defined in the table cover several types of socioeconomic systems. The Alps, for example, have had – historically at least – three types: socialist (Slovenian Alps) heavily subsidized market entities: (EEC – French, Italian, German Alps) and somewhat less subsidized cantonal/communal systems (Swiss, Austrian Alps). This, and the patchwork of local laws, creates much variation not only in living standards but also in funds available for preventive or protective measures to cope with climate change.

This variation may be expected to continue and increase in the future as the socialist countries swing over to a market system and more and more functions are decentralized. Mountain regions in many parts of Europe are asserting ancient independent claims. Ethnic loyalties are strong in mountain regions. Conversely to this centripetal force, there are centrifugal movements uniting local governments across national boundaries, as in the Alpe Adria grouping in the Eastern Alps and Alpe Arge in the Western Alps.

Most of what has been said so far is derived from computer modelling of probable global changes. These models are only as good as the material that is fed into them: GIGO – garbage (or guesses) in and garbage (or gospel) out. Much of the data fed in are uncertain, or incomplete. Nor are the computer models finely enough grained, or dynamic enough, to accommodate to the fast-changing situation of mountains. Another source is analogy – either with past warm or cold periods, or spatially with other regions.[11]

In the present interglacial, about six thousand years ago, there was the so-called Hypsithermal when temperatures were about 1.5°C higher than the period 1950-79. Also in the Middle Ages between about 900 and 1050, there was a warm period – the Medieval Optimum, probably about 1°C higher than the 1950-80 period.

The Medieval Optimum is the best known. Forests extended much farther north and higher on the mountains than they do now. Iceland and Greenland were able to be settled. Vineyards

Figure 3.5: Climate change and socioeconomic factors in European mountains

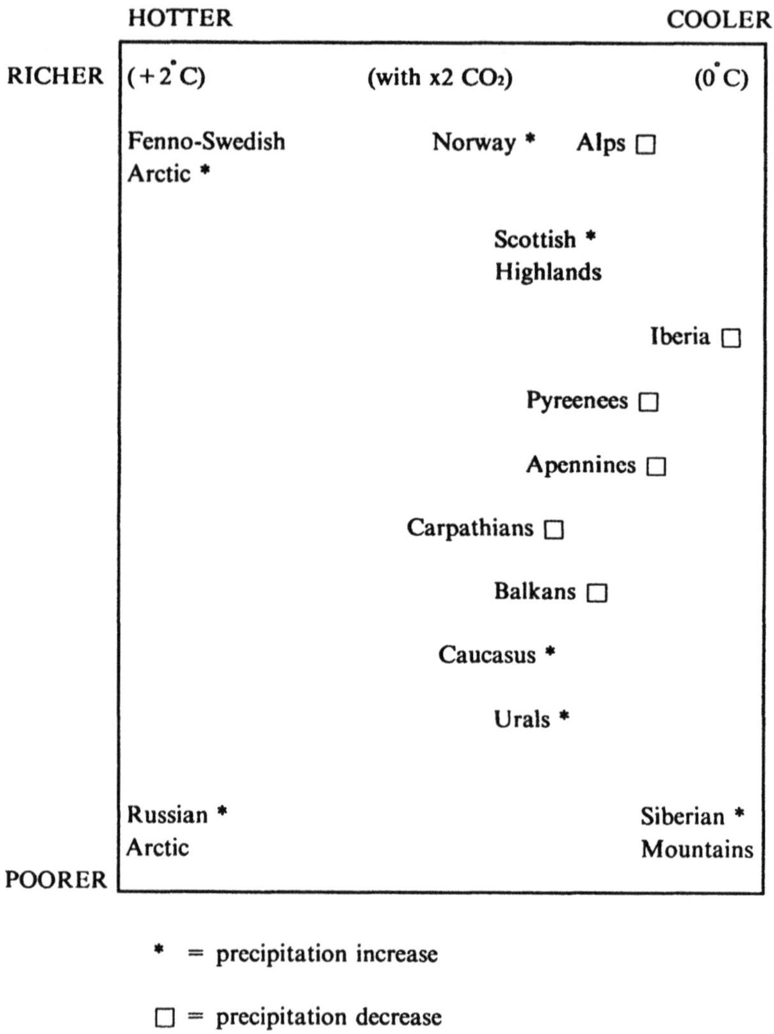

flourished in England. There was considerable prosperity and a healthy population following a series of good harvests. Historians have called this period the Dark Ages because the empires had been eclipsed, but at the grass roots there was much independence, not least in mountain communities. The bully boys of the great powers were away in the Crusades. It is probably no accident that in the twelfth century there was an independence movement in the Alps, the revolt of William Tell, and the establishment of a Helvetic Confederation. By the thirteenth century, the climate had cooled off. There was a series of bad harvests, the population became malnourished, and the Black Death swept over Europe.

Table 3.5: CO_2 and palaeoclimate

Palaeoclimatic warm phases	0°C	Assumed CO_2 content (ppm) (IIASA)	Analogue year
Medieval warm phase (1,000 yrs ago)	+1.0	385–430	2000 (380)[1]
Holocene warm phase (6,000 yrs ago)	+1.5	420–490	
Eem interglacial (120,000 yrs ago)	+2.0 +2.5	460-555 500–610	2025 (420)
Ice-free Arctic Ocean ($12\text{-}2.5 \times 10^6$ yrs ago)	+4.0	630–880	2050 (560)

Brackets refer to estimates made by Soviet scientists.
Sources: Flohn, *Possible Climatic Consequences of a Man-made Global Warming* (Laxenburg: IIASA, 1980).

Climate historians such as Hermann Flohn (see Table 3.5), have interpreted the Medieval Optimum as a shift to the north of the main storm path of rain and cyclonic weather, leaving much of Central Europe under the influence of persistent anticyclonic weather.[12] In many areas there were droughts, but their effect was significant only on food production in the Mediterranean mountains. If the Medieval Optimum is analogous to the present, the warm phase would continue at present levels for one or two hundred years, even with little human intervention. If the greenhouse effect produced higher temperatures (+4°C), the only analogy would be the last interglacial, when the sea level was so high as to make Scandinavia an island and melt all the Arctic ice. This warm, wet world eventually cooled off when the vast amounts of precipitation fell as snow.

By delving deeper into the historical record, it is possible to find analogies of greater warming. Table 3.5 summarizes some of the analogues, both from the last interglacial and earlier. Fossil records for the Pliocene show that the first hominids emerged in this period. The Gribbins have argued that this stage of evolution was triggered by a deteriorating climate.[13] In their thesis, the arrival of the ice stimulated our ancestors to expand their intelligence beyond that of other creatures. Very much warmer conditions, when tropical jungles spread over London, were found much earlier, though there is no certainty about CO_2 levels then. There are traces of ice even from the Miocene, a little over three million years ago. There is no certainty about the cause of the change in the climate. The Gribbins have said that it was possibly the jostling of continents in northern latitudes which cut off the flow of warm water into Arctic regions, so allowing the skin of that ocean to freeze over and thus reflect heat back to space. The evidence seems to indicate a situation where ice was expanding, but temperatures were higher than they are today in Europe, with more precipitation and forests. The Ice Age proper did not begin until the Pleistocene, 1.8 million years ago, when there was also an unexplained reversal in the earths magnetic field.

The contemporary situation is said to be rich in surprises. Mountain environments are especially prone to sudden unexpected changes in weather, or unforeseen events such as earthquakes or avalanches. When IPCC Working Group 1 prepared its scientific assessment, many of the mountain and cryosphere factors were relegated to the surprise box. So, the possibility of huge chips of ice breaking off the Antarctic continent, causing a 6 metre sea-level rise, was largely discounted, though it was documented by satellite pictures. Surprises may be human as much as physical. Wars and revolutions are sudden events with enormous consequences. King Hussein argued at the Second World Climate Conference that a Gulf War, because of extensive oil-field fires, would generate a hundred times more carbon emissions over that region than the upper limits of present scenarios. Sadly, his prediction, which seemed sensationalist at the time, came true in 1991 when the

Gulf environment was all but destroyed by oil slicks, black rain and carpet bombing. One major significance of surprises is that spirals of feedbacks and reactions are set in motion. Unexpected snowfalls or long snow seasons may trigger a vicious circle of poor harvests, malnutrition and bad health, as well as continuing bad weather.

The cryosphere is the most sensitive to climate change of all ecosystems, as well as the most influential.[14] Even the monsoonal rains of India have been correlated inversely with the great extent of the Eurasian snow cover.[15] Over 90 per cent of the earth's freshwater store is in ice, mostly in Antarctica. Ice is a most valuable natural resource, even in such remote places. There are serious studies currently under way of towing Antarctic ice blocks to irrigate the Middle Eastern deserts.[16]

How quickly might the ice melt? There are predictions that there will be no Alpine glaciers in Europe in a hundred years, given the *present* climate.[17] Warming may cause ice surges, leading to sudden calving of polar ice sheets. Sea levels would be raised rapidly and disastrously, and the amount of ice in the ocean might increase the albedo effect. As the ice melted, more moisture would be released and there would be a sudden new ice age, even without increases in cloud cover.

Notes

1. Ozenda, P. and Borel, L., "Possible impacts of a climatic change induced by an increasing greenhouse effect on the Alpine ecosystems", European Conference on Landscape Ecological Impact of Climate Change, Lunteren, The Netherlands, 3–7 December 1989.
2. Professor Turmanidze, Tbilisi, 1990. Personal communication.
3. Singh, T., "Potential impacts of climatic change on forestry,", Canada Committee on Ecological Land Classification Newsletter, no. 17, 1988.
4. Sedjo, R. and Solomon, A., "Climate and Forests" in *Greenhouse Warming: Abatement and Adaptation*, eds. Rosenberg N. et al., (Resources for the Future, 1989).
5. Hinrichsen, D., "Multiple pollutants and forest decline", *Ambio*, vol. 15, no. 5 (1986), pp. 258–65.
6. Sanasilva 1990.
7. Schlaepfer 1990
8. Nilsson, S. "The Price of Pollution: Acid and the Forests of Europe", *Options*, September 1990 (Laxenburg, Austria: IIASA, 1990).

9. Translated from *Journal de Genève*, 26 September 1990.
10. *Journal de Genève*, 13 June 1990.
11. Flohn, H., *Possible Climatic Consequences of a Man-made Global Warming* (Laxenburg, Austria: IIASA, 1980).
12. Ibid.
13. Gribbin, J. and Gribbin, M., *Children of the Ice* (Oxford: Blackwell, 1990).
14. Kuuisto, E., "Snow and Ice", Conference on Climate and Water (Academy of Finland, 1989).
15. Bhanu Kumar, "Eurasian snow cover and forecast of Indian summer monsoon rainfall", *Hydrological Sciences Journal*, vol. 33, no. 5 (1988), pp. 515–25.
16. Peter Clarkson, SCAR. Personal communication.
17. Imbrie, J. and Imbrie, K., *Ice Ages* (Harvard University Press, 1979).

Chapter 4

Climate Change and Mountain Forests

The changes in climate illustrated in previous chapters are expected to affect a wide spectrum of forest processes, forest management operations and, indirectly, socioeconomic development. Today, only a small fraction of the total range of forest responses to climate change is under investigation. We therefore have limited knowledge about the possible negative and positive effects, especially in mountainous areas.

Research today on effects of climate change on forests is concentrated within the areas of growth and forest boundaries. However, even here we face serious problems, the fundamental one being that there is presently no one in the field of ecological theory who can successfully predict the outcome of climate change at any specified time or place. We lack cohesive, tested theories to predict both the behaviour of forest ecosystems in changing environmental situations and the behaviour of species along boundaries of geographic zones.[1]

There are two possible effects on forests of a CO_2 increase and climatic warming. First, there is the direct effect of CO_2 on plant growth. Increases in atmospheric CO_2 could expand the net photosynthesis, resulting in tremendous growth increases. Greenhouse experiments have shown that enhanced CO_2 definitely affects woody seedling growth. Studies carried out in Finland show that a one per cent increase in the atmospheric CO_2 level precipitates an increase of 1.3–3.6 per cent in the basal area of forests.[2] On the other hand, no effects on the growth of established forest stands due to increased CO_2 levels were found in Sweden.[3] It is likely that climate change can disturb the photosynthetic potential of mature trees and reduce the uptake of atmospheric CO_2. Growth losses with enhanced atmospheric CO_2 by mature trees have been demonstrated, and it was found that the photosynthetic capacity tended to decline at increased CO_2 concentrations.[4] The same

Table 4.1: Forest processes influenced by climate change

Regeneration of forests	**Forest depletion**
Insufficient restocking Land availability and quality Seedling production Species selection Site preparation Planting Seedling establishment and survival Genetics and resistance	by fire insects diseases decline and dieback windstorm icing/winterkill/frost flooding/landslides
Growth	**Forest harvesting and transportation**
CO_2 enhancement Temperature/precipitation relations Soil organic matter and nutrient dynamics Soil water	transportability of harvesting machines landslides/flooding fire hazard road/water transportation
Wood	**Wood storage**
Wood characteristics (density, width of tree ring, chemistry composition)	quality changes in storage
	Forest management
	change of silviculture
Forest boundaries	**Socio-economic consequences**
Tree-line Grassland/wetland borders Agricultural displacement	on forest sector and society

Source: Duinker, P. N., Antonovsky, M. Ya. and Solomon, A. M., *Impacts of changes in climate and atmospheric chemistry on northern forest ecosystems and their boundaries: Research Directions*. Working Paper WP-89-14. (Laxenburg, Austria: IIASA, 1989).

occurred with stomatal conductance. Decreased growth can cause the death of trees.

A third possibility would be that different tree species acclimatize to higher CO_2 at different rates. Some species can acclimatize smoothly by changed physiological mechanisms; others will react dramatically. There are indications that forests close to growth boundaries may be the first to respond to changing CO_2 and climate. Such acclimatization indicates that significant changes in boreal or deciduous forest biomass may not be detectable during the first few decades of environmental changes. However, future shifts in forest biomass may show a relative loss of seedlings and saplings compared to mature trees. It has been suggested that a fertilizing effect of increased CO_2 concentrations is more likely to take place in open forest stands with moderate climatic stress.[5] The increased CO_2 concentrations might affect the transpiration and gas exchange, increasing the sensitivity of the forest community structure.[6]

Temperature and soil moisture changes are also likely to have a direct effect on physiological mechanisms. This is especially true of areas where these two factors restrict growth today, such as the boreal forests of North America, the Nordic countries and the USSR, and mountainous areas. It is argued, however, that there will be no physiological effects of increased temperature and soil moisture, unless the degree of mineralization increases proportionately. (This statement is based on conditions in Sweden.[7]) Nutrient limitation and suboptimal climate conditions could limit the extent and rate of reaction by vegetation to all three factors: CO_2 increase, temperature rise and greater precipitation. Opinions differ as to the changed rate of mineralization due to climate change. Climate change may cause a decline in litter quality which could slow down the mineralization rates. This would also slow down forest productivity and create a general decline.[8] Alternatively, the rate of nutrient supply in mineral soils could be accelerated by increased temperature. A temperature rise from 14°C to 17°C could increase the rate of litter breakdown by a third.[9]

Special analyses of the possible effects of climate change have been made in Finland.[10] The growth rate for pine and birch will increase under Scandanavian conditions within the first rotation period. Norway spruce growth response seems to be fairly neutral. The long-run productivity and standing crop of trees could decrease due to enhanced decomposition of litter and humus. On the basis of the Finnish study, it is concluded that the climate change will accelerate the death rate of trees, since the site capacity will be exceeded earlier than it is today. Trees will also mature in a shorter period of time, and annual growth rhythm will be altered. The greatest positive growth changes are projected in the maritime northern part of Europe, whereas growth will level off in the southern areas.[11]

Climate warming is associated with extreme climatic variability of temperature and precipitation. This can cause damage and death to trees and the other components of the forest ecosystem – directly or through the secondary action of pathogens or fire.[12] There is also the risk of increased forest decline due to adoption problems of old genetic strains to climate warming. Not much is known about ecosystematic mechanisms. There are some indications that climate change may encourage the growth of pioneer trees and kill a large number of slow-growing late-successional trees. The disturbance of the annual growth rhythm will make the future structure of forest ecosystems uncertain. Insects, pathogens, diseases and their vectors have their own – often complicated – life-cycles which depend on weather and climatic events in a manner different from that of most trees. Due to climate change, the activities of microbial pathogens and insects will be altered with unknown effects, positive or negative.[13]

The greatest effect of warming on forests is to be expected at the high latitudes. Some observers predict a northerly expansion of forests into previous unforested areas in this region; others argue that this will probably not occur.[14] Forests require centuries to develop, especially in areas with thin soil and a scarce supply of nutrients. The predicted climate change is rapid and will not offer the necessary conditions for forests to develop on new land or remain for long periods.

At southern latitudes, forest land is estimated to lose ground to non-forest vegetation; hence the total production of forest biomass for Europe could take one of the following directions: status quo, decline or expansion. Preliminary results also indicate that rates of migration would be much lower than in the past, due to a limited supply of seeds. This could result in a rapid loss of species from landscapes in the absence of massive reforestation programs.

One of the first problems one encounters when trying to assess the impact of climate change on natural vegetation in Europe is the lack of comprehensive vegetation information. In addition, what is known about the links between climate, ecosystematic mechanisms and human management/influence is not well documented from a scientific point of view. There is a major lack of understanding of the influences of ecophysiological mechanisms. In order to improve knowledge in these fields, IIASA's Biosphere Project[15] has developed a Global Vegetation Model which includes a global database on vegetation and climate. The aim of the model is to simulate the effects of climate change on interactions between climate, soil and vegetation. The ecophysiological processes in the model are under development and, for the moment, rather simple. The model merely takes into account limits for the following parameters: precipitation, temperature of the coldest month, temperature sum (base 50°C), temperature sum (base 0°C), temperature of the warmest month and moisture index based on Priestley and Taylor's potential evapotranspiration ratio.[16] The model works on a grid size of 0.5 longitude, so for the moment, it lacks crucial information on components such as the important nutrient and carbon cycle, soil moisture, and the detailed physiological effects of climate change.[17]

Five different General Circulation Model (GCM) predictions have been made or linked to the Global Vegetation Model to estimate the global climatic response to a doubling of CO_2:

GFDL: Geophysical Fluid Dynamics Laboratory of NOAA, Princeton (Wetherald and Manabe 1986, Manabe and Wetherald 1987).

GFDL-Flux: Geophysical Fluid Dynamics Laboratory – ocean heat flux version (Manabe and Wetherald 1987, Jenne 1989).

GISS: Goddard Institute for Space Studies NASA, Columbia (Hansen et al. 1983)

OSU: Oregon State University, Corvallis (Schlesinger and Zhao 1989)

UKMO: United Kingdom Meteorological Office (Mitchell 1983, Wilson and Mitchell 1987)

Climatic output from these GCM projections has been used as input for the IIASA Global Vegetation Change Model. Special maps have been produced for aggregated biomes for Europe as a whole and for areas within Europe with an elevation level

Table 4.2: Trends in vegetation change according to different scenarios

Scenario /Region	Existing Vegetation	GFDL	GISS	OSU	UKMO
Nordic Countries	Boreal Forests and Cool Conifer are dominating. Temperature Deciduous and Cold Mixed exist. In mountainous areas Boreal forests and Tundra dominate.	Temperate Deciduous and Cool mixed dominate. Boreal forests reduced to small mountainous areas. Tundra has been replaced by Boreal forest and Cool Conifer. Cold Deciduous and Cold Mixed introduced in the northern most part.	Temperate Deciduous, Cool Mixed and Cool Conifer dominate. Boreal forest reduced to mountainous areas. The Tundra area strongly reduced. Cold Mixed introduced in the northern most part.	Temperate Deciduous, Cool Mixed and Cool Conifer dominate Boreal forests reduced to mountainous areas. Tundra area strongly reduced. Cold Mixed and Cold Deciduous introduced in northern most part.	Temperate Deciduous and Cool Mixed dominate. Cool Conifer exists in northern part and mountains. Boreal forest reduced to mountains. Tundra disappeared. Cold Mixed introduced in northern most part.
Northern USSR	Cool Conifer, Boreal Forests and Tundra dominating.	Cool Mixed dominates. Boreal Forests completely replaced by Tundra. Tundra completely replaced by Cool Conifer, Cold Mixed and Cold Deciduous.	Cool Mixed and Cool Conifer dominates Tundra replaced by Boreal forests.	Cool Mixed and Cool Conifer dominates. Boreal forests only in the northern most part and replaced Tundra. Small Tundra areas still exist.	Cool Mixed, Temperate Deciuous and Cool Conifer dominate Boreal forests only exist as replacement of Tundra. Tundra completely replaced. Cold mixed introduced in the northernmost part.
Middle Belt USSR	Cool Mixed dominates.	Western part of Cool Mixed replaced by Temperate Deciduous otherwise no changes.	Western part of Cool Mixed replaced by Temperate Deciduous otherwise no changes.	Western part of Cool Mixed replaced by Temperate Deciduous otherwise no changes.	Temperate Deciduous dominates, remainder consists of unchanged Cool Mixed.
Southern USSR	Warm Grass/Shrub, Cool Grass and Desert dominating. In mountains temperate deciduous, Cool Mixed, Boreal forests and Tundra.	Warm Grass/Shrub and Desert dominating. In mountains Shrub and Temperate deciduous dominate.	Warm Grass/Shrub and Desert dominate. In mountains Temperate Deciduous dominate.	Warm Grass/Shrub and Desert dominate. In mountains Temperate Deciduous dominate.	Warm Grass/Shrub and Desert dominate. In mountains Temperate Deciduous and Shrub dominate.

Table 4.2 continued

Scenario /Region	Existing Vegetation	GFDL	GISS	OSU	UKMO
Denmark Benelux France UK Ireland	Temperate Deciduous dominating with spots of Cold Mixed and broad-leaved Evergreen.	Broad-leaved Evergreen dominating. Scrub introduced in southern France.	Broad-leaved Evergreen dominates. Some areas with Temperature Deciduous left.	Broad-leaved Evergreen and Temperate Deciduous dominate.	Broad-leaved Evergreen dominate. Only minor parts with Temperate Deciduous.
Germany Poland Czechos-lovakia Lower-land Austria and Switzer-land	Western part dominated by Temperate Deciduous Eastern part dominated by Cool Mixed.	Western part unchanged. Eastern aprt changed to Temperate Deciduous.	Western part changed to Broad-leaved Evergreen. Eastern part changed to Temperate Deciduous.	Western part unchanged. Eatern part changed to Temperate Deciduous.	Western part changed to Broad-leaved Evergreen. Eastern part changed to Temperate Deciduous.
Mountain-ous areas France Germany Austria Switzer-land Italy	Dominated by Tundra, Boreal Forests and Cool Mixed.	Tundra replaced by Boreal forests. Former Boreal forests and Cool Mixed replaced Cool Mixed respectively Temperate Deciduous.	Some areas of Tundra left. Most Tundra replaced by Boreal forests Former Boreal forests replaced by Temperate Deciduous.	Some areas of Tundra left. Most Tundra replaced by Boreal forests. Some of former Boreal forests replaced by Cool Conifers and Temperate Deciduous.	Most Tundra replaced by Boreal forests and Cool Conifer. Former Boreal and Cool Mixed replaced by Temperate Deciduous.
Hungary Bulgaria Romania Yugoslavia Albania	Northwestern part Temperate Deciduous Eastern part grass/shrub. Mountainous areas Cool Mixed, Cool Conifer and Boreal forests.	Northwestern and Eastern parts unchanged. In mountains Temperate Deciduous dominating with spots of Cool Mixed.	Northwestern and Eastern parts unchanged. Mountainous areas dominated by Temperate Deciduous with spots of Cool Mixed.	Nothwestern and Eastern parts unchanged. Mountainous areas dominated by Temperate Deciduous with spots of Cool Mixed.	Northwestern part unchanged, Eatern part more shrub. Mountainous areas dominated by Temperate Deciduous while spots of Cool Mixed.

Table 4.2 continued

Scenario /Region	Existing Vegetation	GFDL	GISS	OSU	UKMO
Iberian Peninsular	Southern part dominated by scrub and warm grass/shrub. Northwestern part dominated by Broad-leaved Evergreen. In mountainous areas exist shrub, scrub, Temperate deciduous, broad-leaved Evergreen, Cool Conifers and Temperate Deciduous.	Most of the Northern part changed to scrub, Northern most part to broad-leaved Evergreen. Southern part unchanged.	Most of the Northern part changed to scrub, Northern most part of Temperate Deciduous changed to Broad-leaved Evergreen. Southern part unchanged.	Most of the Northern part changed to scrub, Northern part of Temperate Deciduous changed to broad-leaved Evergreen. Southern part unchanged.	Most of the Northern part changed to srub, Northern part of Temperate Deciduous changed to broad-leaved Evergreen. Southern part unchanged.
Italy (mountains excluded)	Southern part dominated by scrub. Northern part dominated by Temperate Deciduous.	Southern part unchanged. In Northern part large areas of Temperate Deciduous changed to broad-leaved evergreen.	Dominating part unchanged. In Northern part transition of Temperate Deciduous to broad-leaved evergreen.	Dominating part unchanged. In Northern part transition of Temperate Deciduous to broad-leaved evergreen.	Southern part unchanged. Most of the Northern part transformed to broad -leaved evergreen.
Greece Turkey	Greece dominated by scrub and warm grass/shrub. Turkey dominated by warm and cool grass/shrub. In mountains Temperate Deciduous, Cool Conifer, Cool Mixed and Boreal forests.	In Greece more scrub. In Turkey more warm grass/shrub. In mountains former Cool Conifer and Boreal replaced by Cool Mixed forests.	In Greece more scrub. In Turkey more warm grass/shrub. In mount-ains Cool Conifer replaced Temperate Deciduous, Boreal forest replaced by Cool Mixed.	In Greece more scrb. In Turkey more warm grass/shrub. In mountains Cool Conifer and Boreal forests replaced by Cool Mixed and Temperate Deciduous.	In Greece more scrub. In Turkey more warm grass and scrub. In mountains Cool Conifer and Boreal forests replaced by Cool Mixed and Temperate Deciduous.

above 700 metres and 900 metres respectively.[18] It is important to remember that small-scale physical processes are not taken into account in the GCM models. This is an important obstacle for mountainous regions, where ecosystems are small-scale and local climate variations are great.[19]

The GFDL–Flux scenario generates a quite different effect pattern on the vegetation from the other scenarios. In this case the ocean heat flux is statistically incorporated into the GCM model. In the other models, the oceans are represented by a shallow ocean. This means that predetermined sea-surface temperatures are used to regulate the atmospheric circulation. Many uncertainties arise from projections in the GCMs. If oceans are included, the uncertainties will increase. There is a general scientific consensus that it will take about ten years of research to bring the ocean models within today's standard of the General Circulation Models.

The other four scenarios outline how vegetation change may develop as a result of the greenhouse effect in similar ways. The overall trends may be summarized as follows:

- Forest tundra and boreal forests will probably be the vegetation systems in Europe most affected by climate change in terms of area.
- At northern latitudes there will probably be a dominance of temperate deciduous and cool mixed forests, instead of boreal and cool conifer forests.
- In the eastern part of continental Europe, transfer from cool mixed to temperate deciduous forests is foreseen.
- In mountainous areas of continental Europe most of the tundra will be replaced by boreal forests, the former boreal forest and mixed cool forests will be replaced by cool mixed and temperate deciduous.
- At the southern latitudes of Europe, a strong increase in scrub and warm grass/shrub areas is plausible.
- In the mountains at southern latitudes, former cool conifer and boreal forests will be replaced by cool mixed and temperate deciduous forests.

Many further studies of the effects of climate change on the vegetation of Europe's mountains have been carried out. What follows is a brief survey of the findings of some of these studies.

Climate change would involve an upward transition of the vegetation belts. Each existing vegetation belt will take the place of the one immediately preceding it. Vegetation and fauna above the timber line would recede. The flora is estimated to be impoverished. Many of the biotopes would disappear or become fragmented, endangering rare species and making regeneration of vegetation more difficult. This study concludes that it will take several centuries for the forest systems in the alpine regions to stabilize after climate change.[20]

In the long term, higher regions will be more affected than lower regions by increased CO_2 concentrations and increased temperatures. Indirect effects appear to be of much more importance than direct effects for plant processes. The most important factor seems to be the increasing length of the snow-free period. The probable effect will be partial replacement of evergreen dwarf shrubs by deciduous.[21]

In a study of the responses of alpine forest ecosystems to various environmental changes using simulation models, the most striking response to CO_2 increase and climate change occurred at extreme climatic sites – i.e. sites near and above the upper timber line – and on dry sites. As a result of increased temperature, temperature-sensitive trees invaded the high montane sites.[22]

In another study using model simulations (but underlining their strong limitations) the following warnings about effects of increased CO_2 in mountainous regions were stressed:

- a drastic biomass reduction may take place on xeric sites in all altitudinal belts due to increased water stress;
- a significant increase in biomass production may occur on intermediate and mesic sites in the montane, sub-alpine and alpine belt;
- an invasion of deciduous species into existing montane and sub-alpine belts is foreseen;
- a dieback of various coniferous species is foreseen in montane and sub-alpine regions. These species will be replaced by warm-temperate species;
- a migration of coniferous species to high-elevation sites in the alpine belt is foreseen.[23]

The tundra ecosystem in Fennoscandia could completely disappear as a result of climate change. Increased temperature could improve tree growth in (sub)-Arctic and (sub)-alpine zones. Deciduous trees may invade the boreal zone, and coniferous species may move into (sub)-alpine areas.[24]

Various plants from the (sub)-Arctic and (sub)-alpine zones will be among the losers in the event of a climatic change, because plants in these regions have adapted to a more constant environment, have a long life expectancy, live in more constant numbers and have relatively low migration ability.[25]

The potential vertical shift of vegetation belts in European mountains could be in the range of 300–700 metres.[26] Ongoing severe growth declines and dieback of silver fir in the Vosges mountains in France is due to the change in climate.[27]

Although the scientific background material is limited, the general conclusion must be that the mountainous forests of Europe are under severe threat from the predicted climatic changes. The mountain forests are one of the big losers of global vegetation in this changing situation.

The effect on individual species must also be considered. A change in permafrost distributions, shifting nutrient dynamics, expansion of heat-unit accumulation with attendant increases in the length of the growing season, decreased snow persistence periods and loss of winter temperature severity may all contribute to endangering species.[28] Many tree species may benefit from a warming of up to approximately 1°C, but warming above that temperature may result in declined growth and dieback. Warm temperatures can be harmful to populations used to cooler conditions. The net CO_2 influx may be slow on warm and dry days; the phenology may be disrupted; the chilling requirement may not be met; the trees will be unable to handle periods of drought and high temperatures. A temperature increase of 2–3°C does not guarantee an earlier budburst during the spring because there has been decreased chilling and failure to release winter dormancy.[29]

In a study of the historical long-term migrations of beech (*Fagus*) in Europe and North America, the conclusion was that climatic change has been the primary factor controlling migrations and abundance of beech for these two continents. A study of historical migrations of West European oaks led to a similar conclusion.[30]

The life-cycles of different species must be in harmony with the changed weather patterns. Sensitive stages of most species' life-cycles – such as flower production, pollination, seed set, seed germination and seedling establishment – are disturbed by climate change.[31] Losses caused by insects and disease will probably be the first observed effects of climate change in forests. Temperature is the most important climatic factor regulating most mountain insects, and a higher temperature could result in an increased number of insect generations per year. Just one more insect generation could have dramatically increased effects. Insects are better able to adapt to changed climate than trees, and in the short run will have the advantage. Pathogens are even more active than insects under warmer conditions, and increased activity could result in declining forests.[32] Storms and fires constitute another threat to forests, and are likely to increase with climate change. Their frequency, sequence and distribution will change and become more difficult to predict. Forest fire hazards depend on the amount of dead biological material in the forests, and its water content. With higher temperatures and less precipitation, the probabilities of forest fires will rise. An increased windfall rate is also expected.[33]

The biological diversity of the world will be strongly affected, too. Climate may alter population sizes, the distribution of species, the composition of species' habitats and ecosystems, and their geographical extent. It may also accelerate the extinction of different species. Ecosystems and their life communities which are limited to restricted areas are more sensitive to climate change than those which occupy large, undisturbed areas. This also means that in these restricted areas the biodiversity is under an increased threat. The following groups of ecosystems are thought to be especially threatened:[34]

peripheral populations
genetically impoverished species
specialized species
animals
montane and alpine communities
Arctic communities.

There is special concern about mountainous areas. Mountain

ecosystems occupy small areas, have smaller populations and will be more vulnerable than other ecosystems to genetic and environmental pressures.[35] Most of the world's biodiversity is located in small areas which will probably be the most affected by climatic change. In most cases, these restricted areas are surrounded by urban landscapes that many species cannot cross.[36] Protected areas and reserves in the mountainous areas are a case in point. They are located close to the margin, making them even more sensitive to climate change. Different policies for the selection and management of these zones in the future have been proposed. A relocation of the reserves, taking into account the expected effects of climate change and the establishment of new reserves containing a broader ecological variability, has been recommended. This is necessary to minimize future potential losses of species and ecosystems.[37]

There is a strong link between the numbers of species in all ecosystems and the total level of the systems' evapotranspiration. The climate scenarios presented above show that there will be significant changes in evapotranspiration rates, which may lead to a decline of the ecosystems. This decline will disturb the relation between climate and biodiversity. Most of the current discussions about biodiversity centre on the number of species, but there is considerably more involved than this. A species' existence depends on other species. Physiognomic diversity and genetic diversity are critically important, and both are expected to be affected by climate change.[38]

The changing CO_2 rate in the troposphere is thought to be the most fundamental and important factor in the evolution of plant and animal life. The natural evolution of plants and animals has so far followed the composition of the atmosphere, so increased CO_2 concentrations raise concerns about effects on biological evolution as a whole.[39] Yet predicting the microevolutionary consequences of climate change is a complex matter, with so many gaps in understanding that "for even a single species the problem is dauntingly complex".[40]

The future location of vegetation belts is also dependent on animals' response to climate change. At present we have very limited understanding of the complicated interactions between plants, animals and environment. Climate change will not only alter vegetation but will also disturb the animals inhabiting the present vegetation zones, so that the energetics and reproduction possibilities of these animal populations will be affected too. It is not certain that the animals will follow the shift of vegetation, owing to other biotic factors.

Wildlife has biological, social and economic value. The IPCC (1990) emphasizes that "wildlife have recreational and aesthetic value as well as contributing to the preservation of aboriginal lifestyles". In Europe, forestry wildlife contribute to the production of meat, fur, recreation and tourism. Wildlife is the basis for much of the tourism and recreational industry in Europe's mountains. Since the mountain ecosystems in Europe will suffer seriously from climate change, the wildlife in these regions will probably be strongly affected too. There are few quantitative estimates of the effects of climate change upon mountain wildlife. A study in the US Great Basin mountains shows that in the event of a temperature increase of 3°C this region will probably lose 44 per cent of the mammal species, 23 per cent of butterfly species and roughly 20 per cent of bird species. It is also anticipated that the animals' social structure will be influenced; this may lead to genetic changes.[41]

The socioeconomic consequences of climate change on forests and the forestry sector will also be considerable. The following breakdown of climatic effects into regional effects for Europe will serve as a framework for discussion of this aspect:

- in high-latitude areas there will probably be a significant temperature increase in the winter months;
- in the Mediterranean region and in Southern Europe there will probably be a temperature decrease during the winter;
- in Central and Southern Europe there will probably be an increase in the summer temperature;
- in Central and Southern Europe, and in the Mediterranean region there will probably be a decrease in the mean annual precipitation. The opposite will probably occur for Northern Europe;
- in Central Europe, Nordic and Balkan countries and Spain, there will probably be a decrease in the summer precipitation;
- in widespread areas of Central and Southern Europe there will probably be increased evapotranspiration during the summer season.[42]

It has been estimated that with a temperature increase of 3°C, forest production in Sweden will increase by 38 per cent north of Limes Norlandicus, by 16 per cent south of Limes Norlandicus, and by up to 43 per cent between the lines.[43] For other regions of Europe, no quantitative estimates on changed production are available.

The silvicultural implications of climate change for the Nordic countries have been outlined as follows:

- regeneration will improve, but the production of herbs and grass will increase, which requires intensified cleaning of the regeneration sites;
- the demand on tending will increase, especially for deciduous species;
- because climate change will stimulate increased growth, there will be a need for shorter intervals between more intense thinnings;
- because of increased growth, the rotation period will be shortened;
- the mortality rate will probably increase, requiring intensified sanitation fellings;
- silvicultural management will probably have to become more site-specific and intensified;
- improved species/provenance control will be required.[44]

Similar studies for other parts of Europe do not exist. We have therefore tried to estimate the effects on forest production and silviculture for the different regions in Europe in a rough manner in Table 4.2. It is important to point out that these rough estimates are based mainly on knowledge from managed industrial forests. Most of the European mountain forests are not industrial forests but protected forests, parks, etc. with very specific management systems.

Some socioeconomic analysis has been conducted for regions other than Europe. According to one study, the effects of climate change on US forest products industry would be negative.[45] The same was found for the socioeconomic impacts both in the USA and in Canada.[46] A further study found that Canadian forest producers will suffer more than the estimated gains for forest-product consumers.[47] Climate change generates significant cost increases in regions with limited forest-production capacity today, such as Western Europe.[48]

Some aggregated calculations on climate change and development of forest resources in different parts of the world have been made by the US Department of Commerce and are shown in Table 4.4.

Table 4.3: Changed forests production and silvicultural management due to climate

Region	Forest production	Silvicultural implications
Nordic countries	Increased production	Intensified management of all phases of silviculture
Northern USSR	Increased production	Intensified management of all phases of silviculture
Middle-Belt USSR	Increased production in the western part	Intensifed management in the western part
Southern USSR	Decreased production	Intensified protection forest management; intensified sanitation and fire protection
Denmark Benelux France United Kingdom Ireland	Incresed production	Intensified management of all phases of silviculture
Germany Poland Czechoslovakia Lower-Land Austria Switzerland	Increased production	Intensified management of all phases of silviculture
Mountainous areas: France Germany Austria Switzerland Italy	Decreased production	Intensified protection forest management; intensified sanitation and fire protection
Hungary Bulgaria Romania Yugoslavia Albania	Decreased production	Intensified protection forest management; intensified sanitation and fire protection
Iberian Peninsular	Decreased production	Intensified protection forest management; intensified sanitation and fire protection
Italy (except mountains)	Decreased production	Intensified protection forest management; intensified sanitation and fire protection
Greece Turkey	Decreased production	Intensified protection forest management; intensified sanitation and fire protection

Table 4.4: Climate change and development of forest resources

Region	Estimated Temp. Increase C	Steady state[1] change forest land area	Steady state[1] change forest biomass
US	2.5	+5%	+10%
USSR	3.0	+10%	+20%
Australia	2.5	0%	0%
China	2.5	+1%	0%
Brazil	2.0	−2%	−3%
EEC	3.0	+2%	+10%

1. Over 100 year's adjustment to changed temperature and precipitation.
Source: US Department of Commerce (1989).

On the other hand, model analysis of forests in the USA found that at changes more severe than 2°C warming and 12 per cent reduction of precipitation there will be virtual dieback of the forest. These conditions are similar to estimates for the mountainous forests in Central and Southern Europe.[49]

It is clear from the previous discussion that climate change is likely to cause timber-supply fluctuations in Europe as a whole and in its mountainous regions. Very little is known, however, about the socioeconomic impact on rural communities, industrial concerns, markets and trade in forest products, or on governments. The land base for the forest industry may be shifted to other regions as a result of climate change. This could lead to socioeconomic problems in the rural mountainous communities in Europe.[50]

Yet another aspect which must be considered is water. Vegetation interacts both directly and indirectly with the hydrological cycle, so a climate change which affects forest communities will also influence hydrology. Three ways of interactions between vegetation and hydrology have been identified:[51]

vegetation influences:	the amount of precipitation entering the soil the rates of runoff and erosion the water available for evapotranspiration.

Table 4.5 illustrates a number of water-related disturbances due to climatic change.

Since the mountains in Europe are big water "collectors" and

Table 4.5: Water-related disturbance due to climate change

		Land			Rivers and Lakes			
	Wetter (+) Drier (−)	Soil moisture	Groundwater	Surface water	Flow-related uses	Carrier of effluents	Use of water bodies	Use related to water depth
Societal activities concerned		• agriculture • forests • buildings • roads	• rural water supply • local irrigation	• urban activities	• water supply • hydropower • irrigation	• sanitation • industrial waste water disposal	• fishing • recreation	• navigation
Problems encountered	+	• reduced fertility	• waterlogging • building damages	• urban storm runoff • erosion	• failing flow control • floodings		• erosion/ sedimentation • increasing currents • flushing of • lake phosph.	• floodings, inundations
	−	• droughts • crop failures	• drying wells • pumping costs • foundation problems • subsidence		• water deficiencies • reduced hydropower production	• reduced dilution • aeration problems • unsafe for bathing	• changing fish population • reduced water removal	• collapsing navigation systems • reduced traffic
Engineering measures to mitigate	+	• drainage	• drainage	• urban drainage	• flow control		• river training	• increased levee height • reservoirs
	−	• irrigation	• deeper wells		• water transfer schemes	• flow control • waste treatment • aeration	• dredging	• flow control • barrages • sluices • dredging

Source: Brouwer, F. and Falkenmark, M. "Hydrology Water Availability Changes", in Solomon, A.M. and Kauppi, L. (eds). *Towards Ecological Sustainability in Europe. Climate, Water Resources, Soils and Biota*. IIASA, Research Report RR-90-6 (August 1990).

there are to be major changes in forests, we can anticipate major water-related disturbances. These factors will probably cause seasonal water shortages, with many adverse socioeconomic effects requiring local changes in water-storage capacity.[52] Major research is needed on the way changed vegetation patterns will influence water quality.

Anticipated forest decline and changed forest species will also affect the recreational and touristic use of forest land. The pattern of tourism will change in space and time. Winter sports will probably lose some of the most suitable areas. GRCR points out that between 1981 and 1989, there was a decline in the snow cover in Europe of 5.8 per cent in winter, 9.4 per cent in spring and 12.9 per cent in autumn in comparison with the period 1972–80.[53]

The skiing season in Ontario, Canada will be reduced, but opportunities for summer recreations will increase. Winter sports in eastern Canada, the European Alps and the snowy mountains of Australia may no longer have a viable length of ski season within 20–30 years.[54] Warmer winter conditions also give higher risk of avalanches, already the cause of big problems in many winter sport areas. The extent to which the skiing season will be reliable over the Christmas–New Year holidays is a crucial factor for the survival of winter tourism.[55]

Owners of tourist infrastructure in general should be prepared for reduced recreational potential due to forest changes caused by climate change. Tourism is the major source of income for many of the rural communities in the mountainous regions of Europe. The industry should have sufficient understanding of atmospheric sensitivity to plan strategically for the future.[56]

In conclusion, it is clear that our knowledge of the effects of climate change on forests is fragmentary. Most of it is based on quite simplified models correlating vegetation type and forests to gradients of temperature and precipitation, and the ratio of potential evapotranspiration to precipitation. We know that the ecophysiological processes are much more complicated, but for the moment these are the best available tools. These types of models are helpful in illustrating the global distribution of vegetation under a steady state, but they do not take complicated dynamics into account.

Today, the scientific consensus is that a temperature increase of 3°C will occur during the next century. A 1,400-year tree-record of summer temperatures carried out in Fennoscandia has indicated that it will not be possible to detect regional greenhouse signals in the forests until the year 2020.[57] An important contribution to illustrate the effects of climate change has recently been published.[58] This describes the effects on the lakes of the Central Boreal Forests in Canada. Twenty years of climatic, hydrologic and ecological data show that air and lake temperatures have increased by 2°C and the ice-free season has increased by three weeks. This development has resulted in higher evaporation, lower precipitation and less water renewal, leading to higher concentrations of most chemicals. Populations of phytoplankton have multiplied. Wind velocities and transparency have increased. Summer habitats for different organisms have declined.

Perhaps the most convincing data on climate-change effects on forests are historical. During the Middle Ages (800–1200) the mean temperature was 1°C higher than at present in the northern part of Europe. At that time, it was possible to cultivate corn as far as 65°N in Norway and Iceland. This temperature increase also caused substantial shifts in altitudinal and latitudinal tree limits. During the Little Ice Age (1550–1850) mean temperatures were 1–2°C lower than they are today. The land previously colonized by new species was once again abandoned.[59]

Notes

1. Solomon, A.M., "Ecosystem theory required to identify future forest responses to changing CO_2 and climate in ecodynamics", in Wolff, W., Soeder, J., and Drepper, F.R. (eds), *Theoretical Ecology* (Berlin/Heidelberg/New York: Springer-Verlag, 1988).
2. Hari, P. and Arovaara, H., "Detecting CO_2 induced enhancement in the radial increment of trees: evidence from northern timber line", *Scandinavian Journal of Forest Resources* 3 (1988).
3. Linder, S., Proceedings from the Conference on Greenhouse Effects. Uppsala, 19 October 1990, Swedish University of Agricultural Sciences (forthcoming).
4. Kramer, P.J. and Sionit, N., "Effect of increasing carbon dioxide concentration on the physiology and growth of forest trees", in Strands, W.E. and Hoffman, J.S. (eds), *The Greenhouse Effect, Climate Change and U.S. Forests* Washington, D.C.: The Conservation Foundation, (1987); Oechel, W.C., and Strain, B.R., "Native species responses to increased atmospheric carbon dioxide concentration," in Strain, B.R. and Cure, J.D. (eds) *Direct Effects of Increasing Carbon Dioxide on Vegetation* (Washington, D.C.,

1985); Williams, W.E., Garbutt, K., Bazzaz, F.A. and Vitonsek, P.M. "The response of plants to elevated CO_2" *Oecologia*, 69: (1986), pp.454–9.

5. Kienast, F. and Luxmoore, R.J., "Tree-ring analysis and conifer growth responses to increased atmospheric CO_2 levels", *Oecologia* 76 (1986) pp.487–95.
6. Woodward, F.I., "Global change: transplanting plant ecophysiological responses to ecosystems", *Tree*, 5, 9 September 1990.
7. Linder, S., Proceeding from the Conference on Greenhouse Effects; Strain, B.R., and Cure, J.D., "Direct effects of increasing carbon dioxide on vegetation", U.S. Dept. of Energy Report DoE/ER-0238 (1985)
8. Radloff, D.L., "Climate change and America's forests", unpublished manuscript Washington, DC: Forest Service, U.S. Department of Agriculture (June, 1989).
9. Hendrickson, O.Q., "Variation in the C:N ratio of substrate mineralized during forest humus decomposition", *Soil Biology. Biochemistry* 17 (1985) pp.435–40. Kauppi, P. and Posch, M., "A case study of the effects of CO_2-induced climate warming on forest growth and the forest sector", in Parry, M.L., Carter, T.R.and Konijn, N.T. (eds), *The Impact of Climatic Variations on Agriculture Volume I: Assessments in Cool Temperature and Cold Regions* (Dordrecht: Kluwer Academic Publishers, 1987); Binkley, C.S., and Larson, B.C., "Simulated effects of climatic warming on the productivity of managed northern hardwood forests", in Kairiukstis, L., Nilsson, S. and Straszak, A. (eds), *Forest Decline and Reproduction: Regional and Global Consequences*, Working Paper WP-87-75 (Laxenburg, Austria: International Institute for Applied Systems Analysis, 1987); Solomon, A.M. and West, D.C., "Simulating forest responses to expected climate change in eastern North America: applications to decision making in the forest industry", in Strands and Hoffman (eds), *The Greenhouse Effect: Climate Change and U.S. Forests*; Houghton, R.A. and Woodwell, G.M., "Global climatic change", *Scientific American*, vol. 260, no. 4 (1989).
10. Kellomaki, S. Hanninen, H. and Kolstrom, T., "Model computations on the impacts of the climatic change on the productivity and silvicultural management of the forest ecosystem", *Silva Fennica*, vol. 22, no. 4 (1988).
11. Kauppi, P. and Posch, M. (1987).
12. Chen, R.S. and Parry, M.L., "Climate impacts and public policy," (Nairobi: UNEP/IIASA, Nairobi, 1988).
13. Kellomaki *et al* (1988); Smith, U.H., "Changing forest conditions: forest dynamics in a dynamic atmosphere", *Proceedings NE Society of American Foresters Conference* (1987).
14. Houghton, R.A. and Woodwell, G.M. (1989).
15. Solomon, A.M., "Envisaged climate changes could kill boreal forests", International Institute for Applied Systems Analysis, Laxenburg, Austria, *Options*, September 1989.

16. Priestley, C.H.B. and Taylor, R.J., "On the assessment of surface heat flux and evaporation using large-scale parameters", *Monthly Weather Review* 100 (1972), pp. 81–92.
17. Pastor, J. and Post, W.M., "Response of northern forests to CO_2-induced climate change", *Nature,* 334 (July 1988), pp.55–8; Mooney, H.A., Vitousek, P.M. and Matson, P.A., "Exchange of materials between terrestrial ecosystems and the atmosphere", *Science,* 238 (1987), pp.926–32; Bjorkman *et al.* (1987).
18. The IIASA Global Vegetation Change Model is described by: Leemans, R., "Possible changes in natural vegetation patterns due to a global warming", in Hackl, A (ed.) *Der Treibhauseffekt: Das Problem – Megliche* Folgen – Erforderliche Massnahmen Akademic für Umweil und Energie, Laxenburg, Austria, (1989) pp. 105–22; Leemans, R., *IIASA's Climate Data Base,* International Institute or Applied Systems Analysis, Laxenburg, Austria, Options, March 1990; Leemans, R. and Cramer, W.P., *The IIASA Database for Mean Monthly Values of Temperature, Precipitation and Cloudiness on a Global Terrestrial Grid,* Working Paper WP-90-41 (Laxenburg, Austria: International Institute for Applied Systems Analysis, 1990); Monserud, R.A., *Methods for Comparing Global Vegetation Maps,* Working Paper WP-90-40 (Laxenburg, Austria: International Institute for Applied Systems Analysis, 1990).
19. Schlesinger, M.G., "Model projections of the climatic changes induced by increased atmosphere CO_2 climate and geosciences", in Berger, A. (ed.), *NATO Symposium on Climate and Geosciences* (Dordrecht, The Netherlands: Reidel Publishing Company, 1988).
20. Ozenda, P. and Borel, L., "Possible impacts of a climatic change induced by an increasing greenhouse effect on the Alpine ecosystems", European Conference on Landscape Ecological Impact of Climate Change, Lunteren, The Netherlands, 3–7 December, 1989; Mooney *et al.* (1987); Bjorkman *et al* (1987).
21. Korner, C.H., "Response of Alpine vegetation to global climate change", in Rupke, J. and Boer, H.M. (Eds), *Landscape – Ecological Impact of Climatic Change, Alpine Regions (1989).*
22. Kienast, F. and Krauchi, N., "Simulated responses of Alpine forest ecosystems to various environmental changes: application of a forest succession model", in Sjogren, E. (Eds). *Forest of the World: Diversity and Dynamics.* Studies in Plant Ecology 18. Almquist & Wiksell International. Stockholm, Sweden, Uppsala 1989.
23. Kienast, F. "Simulated effects of increasing atmospheric CO_2 on the successional characteristics of Alpine forest ecosystems", European Conference on Landscape Ecological Impact of Climatic Change, 3–7 December 1989.
24. Boer, M.M., Koster, E.A. and Lundberg, H., "Greenhouse impact in Fennoscandia – preliminary findings of a European workshop on the effects of climate change", *Ambio,* vol. 19, no.1 (1990).
25. Koster, E.A., and Lundberg, H. (eds), "Impact analysis of climatic change in the Fennoscandian part of the boreal and sub-Arctic zone", Report prepared for the European Workshop on Interrelated

Bioclimatic and Land Use Changes, Noordwijkeohout University of Utrecht, 1987.

26. Van Huis, J. and Ketner, P., "Climate sensitivity of natural ecosystems in Europe", European Workshop on Interrelated Bioclimatic and Land use changes, The Netherlands, 17–21 October 1987.
27. Becker, M., "The role of climate on present and past vitality of silver fir forests in the Vosges mountains of northeastern France," *Canadian Journal of Forestry Research* 19. 1989.
28. Van Cleve, K. and Dyrness, "Introduction and overview of a multidisciplinary research project: The structure and function of a black spruce (Picea Mariana) forest in relation to other fire-affected taiga ecosystems", *Canadian Journal of Forest Research* 13 (1983), pp. 695–702; Billings, W.D., Peterson, K.M., Luken, J.O. and Mortensen, D.A., "Arctic tundra: A source or sink for atmospheric carbon dioxide in a changing environment?", *Oecologia*, 53 (1982), pp.7–; Dickinson, R.E. and Cicerone, R.J., "Future global warming from atmospheric trace gases", *Nature*, 319 (1986), pp. 109–115.
29. Cannell, M.G.R., Grace, J. and Booth, A., Possible impacts of climatic warming on trees and forests in the United Kingdom: a review. The Journal of the Institute of Chartered Foresters. *Forestry* 62–4 Oxford University Press, 1989; Cannell, M.G.R. and Smith, R.I., "Climate warming, spring budburst and frost damage on trees", *Journal of Applied Ecology*, 23 (1986), pp. 177–191; Murray, M.B., Cannel, M.G.R. and Smith, R.I., "Date of budburst of 15 tree species in Britain following a climatic warming", *Journal of Applied Ecology*, (1989).
30. Huntley, B., Bartlein, P.J. and Prentice, I.C., "Climatic control of the distribution and abundance of beech (Fagust) in Europe and North America", *Journal of Biogeography* 16 (1989), pp. 561–560; Kelly, P.M., Munro, M.A.R., Hughes, M.K. and Goodess, C.M., "Climate and signature years in west European oaks", *Nature* 340, 6 July, 1989.
31. Neilson, R.P., King, G.A., DeVelice, R.L., Lenihan, J., Marks, D., Dolph, J., Campbell, B. and Glick, G., "Sensitivity of ecological landscapes and regions to global climatic change", U.S. Environmental Protection Agency. Environmental Research Laboratory. Corvallis, Oregon, September 1989.
32. Radloff, D.L., "Climate change and America's forests. unpublished manuscript (Washington, D.C.: Forest Service, U.S. Department of Agriculture, June, 1989); Ford, M.J., *The Changing Climate* (London: George Allen & Unwin, 1982); Hedden, R.L., "Global climate change: implications for silviculture and pest management", Fifth Biennial southern Silvicultural Research Conference, Memphis, Tennessee, 1–3 November 1988.
33. Wigley, T.M.L., "Impact of extreme events", *Nature*, 316 (1985), pp. 106–7; Wolffsohn, A., *Fire Control in Tropical Pine Forests*. (Oxford: Commonwealth Forestry Institute, Occasional paper no.17, 1980); Fosberg, M.A., "Climate change and forest fires", in: Berger, A. (ed.), *NATO Symposium on climate and geosciences*

Reidel Publishing Company, (Dordrecht, The Netherlands: 1989); Fried, J.S. and Torn, M.S., "The altered climate – fire model simulating the effects of climate change on the effectiveness of a wildland fire initial attack program", unpublished manuscript. (Berkeley: University of California, 1989).

34. Scheider, S.H., "The greenhouse effect", *Science and Policy Science*, 243 (1989), pp. 771–781; Peters, R.L. and Darling, J.D.S., "The greenhouse effect and nature reserves", *BioScience*, vol. 35 no. 11 (1985), pp. 707–17.
35. Korner, C.H.(1989); Murphy, D.D. and Weiss, S.B., "The effects of climate change on biological diversity in western North America: species losses and mechanisms", in Peters, R.L. and Lovejoy, T.E. (eds), *World Wildlife Fund's Conference on the Consequences of the Greenhouse Effect for Biological Diversity* (New Haven: Yale University Press 1990).
36. Reid, W.V., and Millar, K.R., *Keeping Options Alive. The Scientific Basis for Conserving Biodiversity*, World Resources Institute (1989).
37. Van Huis and Ketner (1987); de Groot, R.S., "*Assessment of Potential Shifts in Europe's Natural Vegetation due to Climatic Change and Some Implications for Nature Conservation*, Working Paper WP-88-105 (Laxenburg, Austria: International Institute for Applied Systems Analysis, November 1988).
38. Neilson *et al.* (1989).
39. Beckmann, G. and Klopries, B., "CO_2 – Austeig in der Troposphäre: ein Kardinal Problem der Menschheit", *Der Lichtbogen*, 208 (July, 1989), pp. 4–31.
40. Holt, R.D., "The microevolutionary consequences of climate change", *Tree*, 5, 9 September, 1990.
41. Cohn, J.P., "Gauging the biological impacts of the greenhouse effect: how might species cope with a warmer world?", *BioScience* vol. 39, no. 3, 1989.
42. Pitovranov, S. and Jager, J., "Climate: Grosswetterlagen Approaches", in: Solomon, A.M. and Kauppi, L. (eds) *Toward Ecological Sustainability in Europe. Climate, Water, Resources, Soils, and Biota*, (Laxenburg, Austria: International Institute for Applied Systems Analysis, Research Report RR-90-6, 1990).
43. Boer *et al.* (1990).
44. Kellomki *et al.* (1988).
45. Regens, J.L., Cubbage, F.W. and Hodges, D.G., "Climate change and U.S. forest markets", in Topping, Jr., J.C. (ed), *Coping with Climate Change – Proceedings of the Second North American Conference on Preparing for Climate Change: co-operative approach* (Washington, DC: The Climate Institute, 1989).
46. Babcock, H.M., "Economic implications to forestry of future climate predictions – climate applications", in: Magiver, D.C., Street, R.B. and Auclair, A.N. (eds), *Forest Renewal and Forest Production. Proceedings of Forest Climate '86* (Ottawa, Ontario: Supply and Services Canada, 1989).

47. Van Kooten, G.C. and Arthur, L.M., "Assessing economic benefits on climate change of Canada's boreal forest", *Canadian Journal of Forest Research*, 19 (1989), pp. 463–70.
48. Binkley, C.S., "A case study of the effects of CO_2-induced climate warming on forest growth and the forest sector: economic effects on the world's forest sector", in: Parry, M.L., Carter, T.R. and Konijn, N.T. (eds), *The Impact of Climatic Variation on Agriculture* Volume I: *Assessments in Cool Temperature and Cold Regions* (Dordrecht/Boston/London: Kluwer Academic Publishers, 1988).
49. Bowes, M.D. and Sedjo, R.A., "Processes for identifying regional influences of and responses to increasing atmospheric CO_2 and climate change", prepared by resources for the Future for the Pacific Northwest Library, U.S. Department of Energy, October 1990.
50. Duinker, P.N., "Climate change and forest management, policy and land use", *Land Use Policy*, April 1990.
51. Mooney *et al.* (1987).
52. Anon. *Statement of Findings and Recommendations from European Workshop on Interrelated Bioclimatic and Land Use Changes* (Bilthoven, The Netherlands: National Institute of Public Health and Environmental Health, 1987).
53. Radloff, D.L. (1989).
54. McBoyle, G. and Wall, G., "Impact of CO_2", *Cahiers de Géographie de Québec*, vol. 31 (1987), pp. 39–50; Galloway, R.W., "The potential impact of climate changes on Australian ski fields", in Pearman, G.I. (eds), *Greenhouse: Planning for Climate Change*, (Melbourne: CSIRO Publications, 1988).
55. Smith, K., "Avalanche hazards: The rising death toll", *Geography*, 73 (1988), pp. 157–58; Smith, K., "Tourism and climate change", *Land Use Policy*, vol. 7, no. 2 (April 1990), pp. 176–80.
56. Duinker, P.N. (1990); Smith, K. (1990).
57. Briffa, K.R., Bartholin, T.S., Eckstein, D., Jones, P.D., Karlen, W., Schweingruber, F.H. and Zetterberg, P., "A 1400-year tree-ring record of summer temperatures in Fennoscandia", *Nature* 346 (2 August 1990).
58. Schindler, D.W., Beaty, K.G., Fee, E.J., Cruikshank, P.R., DeBruyn, E.R., Findlay, D.L., Linsey G.A., Sheaver, J.A.,. Stainton, M.P. and Turner, M.A., "Effects of climatic warming on lakes of the central boreal forest", *Science* vol. 250 (1990), pp. 967–70.
59. Boer *et al* (1990); Bolin, B., Jager, J. and Doos (eds), "The greenhouse effect, climatic change and ecosystems", Scope 29. (Chichester, England: John Wiley & Sons, 1986).

Chapter 5

What Can Be Done?

We have already studied existing emission patterns of greenhouse gases in some detail. In order to outline possible options and strategies to mitigate the effect of these gases on forests and their related ecosystems, we must also look to the future. Future emissions will depend on many individually complex and interrelated factors such as population growth, the expansion of economies, the amount and types of energy consumed and what policies and new technologies are put into place. Many different scenarios have been developed and discussed.[1] The one we have chosen was developed by the US Agency for International Development.[2]

The actual baseline scenario assumes no response to climate change and has a time horizon up to the year 2025. The CFC protocol signed in Montreal in 1987 is, however, taken into account. The purpose of the scenario, (which is presented in Table 5.1) is to make it possible to identify the right options and strategies to reduce greenhouse gases.

By studying current emission patterns and future developments, the following points may be singled out as the most relevant to future strategies:

- emissions will double by the year 2025 in comparison with 1985 if no response to climate change is made;
- emissions from developing countries will increase from a third to half of the global share by 2025 as a result of population increase;
- emission from fossil fuels burning is and will be the largest source of greenhouse gas emissions;
- emission of CFCs will continue to be a central source of contribution to greenhouse gases;
- emissions from agriculture are expected to double by the year 2025, although total emissions from agriculture are less serious than the factors mentioned above;

- Eastern Europe and the USSR will probably be the largest emitters of greenhouse gases owing to increased energy consumption and inefficient energy systems;
- the second largest will probably be South and East Asia owing to increased energy consumption and agricultural production.

These conclusions will be the basis for the following discussion of strategies for mitigation.

A possible balance between energy supply and demand for a sustainable world has been analysed, and it is argued that it is possible to formulate energy strategies which are both technically and economically viable for solving global problems such as CO_2 emissions.[3] The overall solution is an end-use strategy emphasizing energy efficiency in both developed and developing countries. For developing countries it means a transition from fuelwood to the use of modern solid, liquid and gaseous fuels. For both developing and industrialized countries it means increased electrification. Biomass acts as an important swing energy source in the analysis.

Changing the fuels used is an important option for the reduction of CO_2 emissions, moving away from fuels with a high CO_2 emission coefficient. Table 5.2 shows emission coefficients for different energy carriers.

Table 5.2: CO_2 emission coefficients (kg CO_2/GJ[a]) for various energy carriers

Energy carriers	CO_2 emission coefficients
Coal	94
Oil	73[b]
Gas	56
Uranium	0
Renewables	0

[a] Gigal Joule primary energy at lower heating value, combustion of fuels, indirect emissions excluded.

[b] CO_2 emissions from refineries are included.

Source: Blok, K., Bijlsma, J., Fockens, S. and Okken, P. A., *CO_2 emission coefficients for fuels and energy carriers in the Netherlands* (ESC-WR-88-12) (Petten, The Netherlands: Utrecht University, 1988).

Table 5.1 Emissions from major greenhouse-gas-producing activities. Baseline case with no responces to climate change. Expressed in million metric tons on CO_2 equivalents.

	CO2 from energy[a]		CO_2 from deforestation		CFCs[b]	
	1985	2025	1985	2025	1985	2025
United States	5,100	6,200-7,700	-	-	1,350	1,675
OECD Europe/Canada	2,900	4,000-5,100	-	-	1,275	2,340
OECD Pacific	1,100	1,500-2,200	-	-	510	935
Eastern Europe/USSR	5,100	8,400-10,300	-	-	710	1,735
Poland	436	935	-	-	-	-
Centrally Planned Asia	1,900	3,700-7,300	220	250-660	90	650
China	1,844	6,303	-	-	-	-
Middle East	400	1,800-2,600	-	-	40	275
Africa	700	700-2,900	620	660-1,210	200	1,400
Zaire	-	-	-	-	-	-
Latin America	700	1,500-3,300	1,280	1,950-3,040	140	990
Brazil	150	473	-	-	-	-
Mexico	249	730	-	-	-	-
Rest of Central America	-	-	-	-	-	-
South & East Asia	1,100	1,800-5,900	550	630-1,690	270	1,890
India	359	2,273	-	-	-	-
Indonesia	81	517	-	-	-	-
Pakistan	-	-	-	-	-	-
Philippines	-	-	-	-	-	-
Thailand	-	-	-	-	-	-
World	19,100	29,700-45,500	2,670	3,500-6,600	4,580	11,890

Note: Several major emissions-producing activities are accounted for, but many of the smaller source categories are not included. Estimates for 2025 are taken from 2030 Emissions scenario, except for the CO_2 estimates from energy, which also include the results from the EIS/AFOS Reference case.

[a]The CO2 emissions itemized in this table do not include emissions from cement production.
[b]Include CFC-11, CFC-12, CFC-113, HCFC-22, CCl_4, CH_3CCl_3, and Halon 1301.

Source: US Agency for International Development (1990).

Table 5.1 con'd

CH_4 from rice		CH_4 from animals		N_2O from fertilizer		Total		Tons per capita	
1985	2025	1985	2025	1985	2025	1985	2025	1985	2025
17	21	206	267	128	156	6,801	8,319-9,819	28	29-34
15	15	311	397	136	268	4,637	7,020-8,120	11	14-17
38	27	97	168	11	16	1,756	2,646-3,346	12	16-20
34	38	277	344	142	266	6,263	10-783-12,683	15	21-25
-	-	-	-	-	-				
647	687	145	279	139	230	3,168	5,796-9,806	3	3-6
561	571	134	254	129	211				
13	17	21	59	11	27	485	2,178-2,978	4	8-11
55	116	107	277	25	95	1,707	3,248-5,998	3	2-4
-	-	-	-	-	-				
124	147	200	441	33	93	2,477	5,121-8,011	6	7-11
86	78	57	92	8	14				
2	4	23	63	13	46				
34	63	59	176	12	28				
1,331	2,056	212	384	99	427	3,562	7,187-12,347	2	3-5
653	1,084	111	189	53	144				
153	302	10	23	12	155				
25	34	34	61	13	75				
-	-	-	-	-	-				
-	-	-	-	-	-				
2,297	3,124	1,576	2,617	723	1,580	30,946	52,411-71,311	6	6-9

Energy Efficiency

The Stockholm Environment Institute recently published an optimistic report on the possibility of reducing world energy consumption significantly through increased energy efficiency. A feasible goal could be a 50 per cent reduction in the growth of total worldwide usage between 1986 and 2030. The report concludes that this could be achieved without harming the global economy, but it stresses that there are substantial obstacles in the way of adopting the necessary measures. Such efficiency improvements will not happen unless important new policies are drawn up and enforced.[4]

Energy efficiency or conservation includes two major facets: the improvement of the energy efficiency of our activities and structural change promoting less energy-intensive activities. The principal sectors where energy efficiency could be improved are:

household and services;
industry;
transport;
power generation.

Table 5.3 shows a carbon dioxide improvement potential index for the Netherlands indicating that households and industry have the best potential for energy efficiency/conservation in industrialized countries.[5]

Table 5.3: Carbon dioxide improvement potential index

CDIP Index	
Household	100
Industry	94
Transport	52
Power generation	52

Source: Diepstraater und Van Gool (1989).

However, the authors underline the fact that energy efficiency is hard to tackle because of the large numbers of small users involved. It is thought that it might be possible to improve the energy efficiency of the household sector in the Netherlands by one per cent per year. The transport sector faces similar problems. Another study has shown that with technical improvements to vehicles, changed occupancy ratio of the means of transport, changed volume development (passengers and ton kilometres) and alternative fuels, it would be possible to reduce energy consumption in transport by 25–45 per cent.[6] Energy efficiency in industry could also be improved by 2–3 per cent a year through conservation. The power-generation sector has untapped possibilities of co-generation, where electricity and useful heat are produced together in a common thermal power plant.

Nevertheless, these analyses based on the Netherlands show that, because of increased demand, even with maximal energy conservation there will be only a slight reduction in total energy consumption, and more drastic means than conservation will be required in a growing economy to achieve stabilization of carbon dioxide emissions. The following general conclusions may be drawn:

- there is theoretical potential for improved energy efficiency;
- a number of obstacles lie in the way of its implementation:
  - lack of knowledge on efficient technologies
  - lack of care
  - demand for high return on investments
  - income from investment usually going to a third party.

Other studies reach similar conclusions. In the USA, complicated market barriers are just one of the hindrances to achieving high energy-use efficiency, yet research shows that improved energy efficiency and conservation are of vital importance in attempting to palliate climatic change. The US Agency for International Development (1990) argues that improved end-use energy efficiency has the greatest potential for slowing the growth of CO_2 and other greenhouse gases in the near future. A recent study showed an average efficiency rate in the USA of 25 per cent – proof, according to the author, of the high potential for energy efficiency/conservation.[7]

Alternative Energy

From Table 5.2, it may be seen that a shift from coal and oil to renewable sources would greatly improve emission conditions. A large number of renewable sources are also available such as solar energy, ocean thermal energy, wind energy, photovoltaics, alcohol-based fuels and biomass energy.

Forests may be used not only as a carbon sink (to be discussed below) but also as fuel and for the replacement of fuels with high CO_2 emission coefficients. To make major improvements to emission conditions, however, huge areas would be required. If half the world energy consumption were supplied by forest biomass, 600 million hectares would have to be exploited sustainably. This means extremely large-scale activities in order to influence the overall emission pattern by the introduction of forest biomass as fuel.[8] It has been calculated that 465 million hectares of new forest plantations would be needed to offset the estimated annual increase of 2.9 billion tons of free carbon. The wood energy potential from these plantations would be high. One year's production would be greater than the energy generated in the US coal-fired power stations in two years.[9] So the substitution of wood for fossil fuel has advantages in terms of the carbon cycle, if the overall stock of biomass is maintained. But today, wood is not regarded as an important source of energy. It is regarded as a "poor man's energy" and the market does not consider wood as competitive with fossil fuel energy. Thus what counts is how financially competitive this fuel would be in comparison with other sources.

Hydroelectric power has great potential.[10] Global production was 2112 TWh/year in 1988. The following hydroelectric potentials have been calculated:

theoretical hydroelectric potential	44,000 TWh/year;
technically accessible potential	19,000 TWh/year;
economic accessible potential	9,700 TWh/year.

The countries with the greatest potentials are the USSR (3,900 TWh/year), Argentina (2,400 TWh/year) and China (1,900 TWh /year). Increasing pressure from environmental groups, however, makes it unlikely that funding organizations will support large hydroprogrammes in the future. On the other hand, there seems to be room for small hydroplants.[11]

The literature is full of promising reports about other alternative energy sources such as solar, wind, ocean, and so on. These promising features will not be repeated here. We shall merely describe what has become of the development of alternative energy sources in the USA since the oil crisis in the early 1970s.[12]

- Solar heating and cooling systems: Installed in fewer than one million of the nation's 80 million households and rapidly losing momentum. Roughly 100 of 120 manufacturers recently went out of business, resulting in a loss of 36,000 jobs.
- Ocean thermal energy conservation: No systems in commercial operation.
- Tidal power: No systems in commercial operation.
- Electricity from industrial processes' waste heat: Between 1978 and 1988, 2800 co-generation facilities were installed; capacity increased by 4 per cent.
- Wind energy: Accounts for less than one-third of 1 per cent of installed US generating capacity.
- Electricity and steam heat from geothermal resources: Installed generated capacity is comparable with totals for wind.
- Small-scale hydropower: About 1,500 units installed, enlarging the national generating capacity by 1 per cent.
- Wood-fired power plants: Four plants installed; total capacity is less than one-thirtieth of small-scale hydropower.
- Solar thermal generators: Eight systems installed, with a total capacity smaller than the wood-fired units.
- Photovoltaics: Worldwide sales in 1986 were 23 megawatts. This corresponds to about half the capacity of the smallest US wood-fired power plant.
- Alcohol-based fuels: In 1988 about 1,000 alcohol-fueled vehicles out of 130 million in the USA.
- Synthetic fuels: Energy-supply less than one-sixth the output of small hydroelectric units (about 1 per cent of the total generating capacity in the USA).
- Fusion power: No systems near commercial operation.

Gas has a lower CO_2 emission coefficient than coal and oil fuels, but the emission coefficient is still high, as we can see from Table 5.4. Natural gas produces significant quantities of the greenhouse gases NO_X and CO_2. During production, transportation and burning, the gas also leaks methane. A number of natural gas-intensive scenarios have been analysed, and the methane leaks found to be significant.[14] These scenarios increased the greenhouse effect by 10 per cent in comparison with the utilization rate of gas today, but it was also found that it might be possible to replace about 20 per cent of coal by gas and still achieve improved emission conditions.

Table 5.4: Estimate of recoverable resources of natural gas (expressed in $10^{12}m^3$)

Region	Cumulative production	Proved reserves	Undis-covered resources	Estimated total recoverable
North America	20.0	8.4	28–36	36–44
South America	3.4	4.9–5.2	7.7–10.8	12.6–16
Western Europe	2.2	5.4–5.6	4.2–6.0	9.6–12
USSR and E Europe	5.9	15.5	13–25	28–40
Africa	1.0	5.7–6.3	7.5–28	13–34
Middle East	2.5	17–24	30–37	47.61
Asia/Oceania	0.5	4.5–6.0	29–36	33–42
Antarctica	0	0	2.3–4.0	2.3–4.0
WORLD TOTAL	35.5	61–71	122–183	182–253
(in 10^{15}gC)	(19.2)	(33–38)	(66–99)	(99–137)

Source: Rotty and Masters (1985)[13]

The threat of climatic change, together with acid rain and other air pollutants, have led to arguments for a broader acceptance of nuclear power.[15] In model analyses, coal-burning is completely replaced by nuclear power within forty years. In order to achieve this a new, large nuclear plant would have to be built every three days from now until 2025. In addition, the CO_2 would continue to grow by about 25 per cent from now until the year 2,000 and decline to today's level around the year 2025.[16] So there are many doubts about how great a contribution nuclear power would make in mitigating climate change. There is considerable popular – and therefore political – opposition to the nuclear option. This is especially true when we take into account the amount of nuclear plants required, increasing the risk of nuclear accidents. The costs – about US$145 billion per year – would be tremendous.[17] It has been argued that each dollar invested in energy efficiency in the USA can displace about seven times more carbon than each dollar invested in nuclear power.[18]

In the political and environmental debate (see Chapter 6) there have been calls for a 30 per cent CO_2 reduction over 1988 figures.[19] At the Second World Climate Conference it was concluded that emissions of greenhouse gases must be reduced by

60 per cent if the atmospheric concentration of these greenhouse gases is to be halted. To achieve this objective, a dramatic change in the consumption pattern of energy is required. It may be concluded that:

- Energy efficiency has great potential, but given the many obstacles in its path it will not have much effect on emission patterns.
- The alternative source of wood for energy is potentially good, but the required scale of operations and the poor economic competitiveness with other sources make it difficult to foresee any introduction of large-scale operations.
- Hydropower shows potential, but environmental restrictions will limit the practical impact of this source.
- Other alternative sources like solar, ocean, wind, etc., show high theoretical potential, but history has taught us that practical implementation will probably be limited.
- Natural gases have strong potential, but are not a solution to the problem owing to significant leaks of methane.
- Nuclear power is doubtful owing to the scale of operations required, risks of accident and high costs.

After scrutinizing about thirty-five different scenarios for future energy consumption and related CO_2 emissions, as well as producing his own, this is what Sinyak concludes about energy patterns up to the year 2060[20]:

- there will be an increased demand for energy;
- energy efficiency will assume growing importance;
- nuclear development will be severely limited during the next few decades;
- renewable sources will contribute little to the global energy balance owing to unfavorable economics;
- fossil fuels will remain the leading global energy source.

He says: "in spite of many radical measures that could be taken [in the energy sector]. . . it is unlikely that increasing CO_2 concentrations can be avoided."

The Second World Climate Conference (1990) concluded that technically feasible and cost-effective means to reduce CO_2 emissions exist in all countries. The measures include improved efficiency of energy use, and the use of alternative fuels and energy sources. But the conference also concluded that these measures are "slowed down and sometimes prevented by a host of barriers".

Very long time lags are involved in reducing emissions through changed energy patterns[21]:

- political time lags: the time needed for political decisions on countermeasures to be taken and enforced;
- technological time lags: the time needed for changes in energy infrastructure to be made;
- atmospheric time lags: the long residence time of greenhouse gases;
- ecological time lags: the time needed for ecosystems to respond to changed emissions.

These time lags add up to a number of decades before the effects of countermeasures in the form of changed energy systems could be identified.

The German Bundestag concludes from a comprehensive study that "there is growing evidence suggesting that the problems raised by the greenhouse effect are so substantial that the structure of the global energy system should be subjected to a fundamental review, involving the readiness to induce major changes."[22]

A study of possible energy policies and the climate change for China, India, Japan, the UK, the USA and the USSR concludes that even in the most optimistic scenario, with each country adopting policies for high energy efficiency and a fuel switch to lower carbon emissions, the emissions of the six countries in the year 2030 will be higher than they were in 1988.[23] The overall conclusion must be that the prevention of the greenhouse effect through changed energy consumption will in reality be very difficult to achieve without extreme and extraordinary actions.

Chlorofluorocarbons and Halons

CFCs and halons will remain a major problem in spite of the Montreal Protocol – partly because developing countries were allowed to delay restrictions against CFCs, but also because of the Protocol's scope: emissions of CFC-11, CFC-12, CFC-14 and CFC-15 will be reduced. However, the reduction of these gases will be offset by the introduction of new gases which are also greenhouse gases. An improved protocol applying to all countries and implementing immediate action is required and should include the entire index of CFCs and halons which contribute to the greenhouse effect.

Tropical Deforestation

The role of deforestation and reforestation in mitigating the effects of climate change must be examined – here, tropical rather than mountain forests are the principal focus. For at least twenty years tropical deforestation has been recognized as a serious problem, but little effective action has been taken to minimize or control it. Deforestation results in a loss of food, energy, shelter and other resources for local inhabitants. It also has a number of environmental effects, the most serious of which is its contribution to greenhouse gases.

The problem of tropical deforestation has worsened dramatically over the past ten years: figures suggest that it has increased by about 100 per cent.[24] The development of deforestation is illustrated in Table 5.5.

Table 5.5: Rate of deforestation (expressed in million ha per year)

Year	1980 (Lanly/FAO)	1989 (Meyers)	1990 (WRI)	1990 (FAO)
	Closed/Total Trop. Forest	CTF/TTF	CTF/TTF	CTF/TTE
Deforestation in million ha per year	7.3/11.3	13.9/22.0	n.a./16–20	n.a./17.0

According to the latest estimates, the largest losses of tropical forests have taken place in Brazil, India, Indonesia, Burma, Vietnam, Thailand, the Philippines and Costa Rica. At present there are about 0.8 billion hectares of tropical forests left. Newly published estimates on the rate of deforestation support earlier scenarios claiming a probable loss of 45–50 per cent of the existing tropical forest by the year 2025 if special countermeasures are not introduced.[25]

Over the last ten years there has been considerable discussion about deforestation's contribution to the greenhouse effect. The most important gases released by tropical deforestation are CO_2, CO_4 and CO. Carbon monoxide is not a greenhouse gas but indirectly affects the concentrations of other greenhouse gases such as CH_4.[26]

Future carbon emissions from tropical forests may be similar in size to current emissions from fossil fuel combustion. The current emissions of CO_2, CH_4 and N_2O, from both the global and the tropical deforestation points of view, have been calculated. The results show that tropical deforestation contributes about 25 per cent of radiative emissions today. Based on new estimates of the rate of tropical deforestation, it is thought that in the future, emissions from this process may grow more rapidly than those from the worldwide combustion of fossil fuels.[27]

The process of tropical deforestation is tremendously complicated.[28] The forests have fallen victim to the general lack of socioeconomic progress in the developing countries. The driving force behind tropical deforestation is socioeconomic, and socioeconomic factors are area-specific for each region of the tropical forest. They also change over time: one can identify an evolutionary process in the rate of deforestation. It may be said that deforestation has been – and still is to some extent – a necessary element of socioeconomic development.[29]

To halt tropical deforestation is an overwhelming task. It requires, as a starting point, a general improvement of the socioeconomic conditions in the tropical forest regions. Experienced development experts fear that the necessary development will not be achieved until a very long time after the extinction of tropical forests.

Reforestation/Afforestation

A wider, global perspective on reforestation is needed. Forests are regarded as the only effective terrestrial source for the fixation of CO_2 and the sequestration of carbon. It has been estimated that forest ecosystems account for a reservoir of over 1,200 billion tons of carbon which is higher than in the atmosphere (700 billion tons).[30] According to another study, 400–800 billion tons are stored in the world's 4 billion hectares of forest.[31]

The IPCC Working Group I made calculations based on a fixation rate of 5 tons carbon per hectare and year and found that a reforestation of 400 million hectares at a rate of 10 million per year from now until 2030 would result in a sequestration of one billion tons of carbon per year over a period of forty to a hundred years. Other calculations give the following results:

- 200 million hectares of new forest will store around 1 billion tons of carbon per year;[32]
- 120 million hectares of fast-growing tropical plantations would sequester 800 million tons of carbon per year;[33]
- 465 million hectares of plantations are required to sequester the annual increment of 2.9 billion tons of free carbon;[34]
- 800 million hectares are required to sequester 3 billion tons of carbon.[35]

Such high production figures may be misleading; and for long-term purposes covering large areas, much lower production values should perhaps be considered.[36] It should also be emphasized that the sequestering capacity in these plantation forests will be sustainable only if the overall greater stock of forest biomass is maintained. If the wood is cut and decomposed, the carbon fixation is only short-term.

There are reasons other than carbon sequestration for the need for increased reforestation; and these are summarized in Table 5.6. In the table, the required reforestation is expressed in equivalent areas of closed plantations. These plantantions, for purposes other than seuqestration, will store about 700–800 million tons of carbon per year, but over shorter periods in comparison with the earlier calculations.

Is reforestation of this scale realistic? The areas required correspond to about 75 per cent of the non-forested land of the United States.[37] During the early 1980s, the yearly rate of reforestation was roughly 10 million hectares; at the end of the 1980s it was estimated at 15 million hectares.[38] The total reforestation in tropical areas in 1985 was 1.7 million hectares.[39] At this rate, given the requirement of plantations, it would, according to one calculation, take at least thirty years to establish them.[40] In addition, the literature illustrates numerous examples of failed reforestation projects.[41] In the light of these experiences it seems unrealistic to be optimistic about large-scale plantations. Rural people cannot be expected to contribute to programmes for long-term benefits, and carbon fixation in itself could never be an objective for reforestation.[42]

Table 5.6: Estimates on needs for reforestation (million hectares of equivalent closed plantations)

Areas needed to cover deficits in wood production	
Fuelwood deficits in 2000	55–100
Industrial wood deficits in 2000	20–40
Degraded areas needing protective tree cover	
Degraded upland watersheds[1]	44–80
Desertified (semi-)arid lands[2]	±50
Wasted fallow lands	23–40
	Total 142–310

1. Assumed 50% tree cover.
2. Assumed 15% tree cover.

Source: Wiersum (1989)

The costs of the plantations would also be tremendous. Land procurement and plantation of 465 million hectares would probably cost no less than US $500 billion: about 13 per cent of the US Gross National Product. The disposal of wood from the plantations would also be expensive. To maintain its carbon sequestering capacity, additional costs for harvesting, transportation and preserving would be at least US$3.1 trillion for a thirty-year period. If the wood from the plantation were to be used for industrial products, it would severely disturb world markets and have negative effects. In this case, the long-term sequestering capacity would also be lost.[43]

Another way of increasing the sequestering capacity of the forest would be to increase production of biomass by intensified management. To sequester all the excess atmospheric carbon, a dramatic increase of biomass production is required. The stemwood production would have to increase by 2.5 cubic metres per hectare per year in all the world's 3 billion hectares of closed forests.[44] Such an increase may be compared with the average growth in some regions:

USA: 3.15 m^3/ha/year
USSR: 1.35 m^3/ha/year
Western Europe: 4.10 m^3/ha/year.

Based on this comparison, it seems unrealistic to expect any dramatic change of the forests' sequestering capacity by improved forest management.[45]

Another option would be to change the use of existing forest resources to sequestration only; this would mean reducing industrial wood harvests. It has been calculated that a worldwide no-harvest policy would probably increase the sequestering capacity of forests to meet only one-tenth of excess atmospheric CO_2.[46]

In earlier sections, the possible effects on existing forests of increased CO_2 concentrations, temperature and precipitation were discussed. We saw that the potential effects were still uncertain, but one possibility is natural forest expansion. Natural forest expansion of closed forests of about 3.5 billion hectares would be required to sequester the excess carbon, representing more than twice the world's existing closed forests (3 billion hectares).[47] It seems unrealistic, therefore, to expect any dramatic changes in the forest carbon sequestering capacity by natural expansion.

From the preceding discussion of deforestation and reforestation/afforestation, it may be concluded that:

- the rate of deforestation will not be halted until real socio-economic development has been achieved in the tropical regions;
- this will probably be achieved only after the tropical forests have disappeared;
- theoretically, the forests could sequester the excess carbon perfectly with increased production, mainly through new plantations;
- in reality, though, this potential seems greatly reduced because of:

    the scale of plantations required
    the long time needed to establish them
    failures in former large-scale plantations
    overwhelming costs.

Realistically, therefore, the forests cannot easily make a major contribution to solving the greenhouse problem. The US Agency for International Development seems to share this opinion, warning against the feasibility of either a dramatic reduction in deforestation or a major increase in reforestation. Its conclusion is that social, economic, political and ethical constraints are too great for any dramatic cut in deforestation or any increase in large-scale plantations to be expected.[48]

None of this, however, should detract from efforts to reforest mountain areas. Mountain forests play a significant role in erosion prevention and watershed management, quite apart from any effects they may have on climate (and microclimates). Mountain and boreal forests make up a significant portion of the world area. In industrialized countries where there is forest dieback, there is a need not only for replacement planting but also for improved management techniques. Last but not least, planting programmes can be an important means of mobilizing and educating the community to prepare for the effects of climate change.

Agriculture

Reforestation is at best only a partial means of combating the effects of climate change; other sectoral actions must be included in any forestry projects. Agriculture contributes significantly to increasing greenhouse concentrations.[49] Specific actions within agriculture to reduce greenhouse gas emissions are difficult to introduce and control. The US Agency for International Development suggests the following measures:

- agricultural production involving biomass burning should change over to alternative practices such as sustainable agriculture and the use of crop residues;
- more efficient management of livestock waste could reduce methane emissions by 25–75 per cent per unit produced;
- use of improved fertilizer formulations and application technologies would reduce nitrous oxide emissions;
- marginal land for annual cropping should be shifted to perennial production or pastoral land uses;
- to improve management of water regimes, cultivars should be developed. This is especially important in rice production.

Others are doubtful, however, that anything can be done to minimize the continuing accumulation of methane and nitrous oxide from agriculture, given that the underlying factors are demographic and very difficult to tackle.[50]

So far we have discussed the possibilities of influencing the major sources of greenhouse gas emissions: energy, CFCs, deforestation and agriculture. We have seen that the underlying factor of the future emission pattern would be population growth, generating special problems, especially, in developing countries. The USSR/ Eastern Europe region was identified as being a major future contributor of emissions.

Population

From the scenario on future emissions of greenhouse gases it can be seen that demography is a crucial factor. An exploding population is to blame for most releases of these gases. If we are to have any hope of controlling the emissions, it has been argued, "the world population cannot be allowed to continue along its present trajectory – it must be stabilized and soon."[51] The upward population trend plays a decisive role in the greenhouse problem.[52] World population is expected to reach 9 billion by the year 2050, with developing countries making up three-quarters of humankind. The foreseeable consequences will be that:

- increased population in developing countries will shift the principal activity away from agriculture to fuel-intensive industry and transportation;
- population growth and rising income will result in an increased number of cattle and greater expanse of rice fields, leading to increased emissions of methane;
- the population growth will lead to an explosive growth of cities and towns and a consequent move from biomass fuels towards fossil fuels.[53]

Elsewhere the population problem is described as "population pressures", or "human-induced emissions".[54] IPCC's conclusion document (1990) states that "a rapidly increasing world population will mean greater demands for resources in general and energy in particular." IPCC's Special Committee on the Participation of Developing Countries (1990) fixed its attention on insufficient information and communication, limited human resources, institutional difficulties and limited financial resources. The ISSC/UNESCO Framework for Research on the Human Dimensions of Global Environmental Change (1990) states that human beings now significantly affect many of the systems of the biosphere and the geosphere, although nothing is said in the document about the influence of the globe's population size on sustainable environment.

It is obvious, though, that if we are to contain the greenhouse problem and many other environmental problems, population dynamics must be in harmony with the ecological support system. At present, this does not seem to be the case, and the problem is not sufficiently discussed because of sociopolitical and ethical sensitivities.

On the other hand, it should be recognized that it is not simply population numbers which determine emissions and pollution. The average American is said to consume forty times more energy than the average Ethiopian. It has been suggested that as important as family planning is a greater voluntary simplicity in lifestyles, especially in rich countries, along with the elimination of waste and unnecessary luxuries. Such policies have proved unpopular in the past, but they may be forced on populations by events.

Developing Countries and the USSR/Eastern European region

We have seen that the developing countries and the USSR/Eastern European region may in future be the largest emitters of greenhouse gases, because of high population growth in many poorer undeveloped regions. The following points highlight the scale of problems facing these countries:[55]

- Debt Crisis: The total debt for developing countries was more than US$1 trillion in 1987 (WRI, 1990). The Soviet external debt was $50 billion in 1989.[56]
- Balance-of-payment crisis: During the 1970s, the net capital flow was from the OECD to the developing areas. Since 1981, net capital flows have reversed. In 1988 this flow was US$50 billion. The USSR/Eastern European region faces a similar dilemma.
- Capital for investment crisis: Debtor countries must export in order to earn hard currency to pay interest on debts. In the case of the USSR, capital is needed for the importation of basic food supplies, so no capital is available for new investments.

- Limited export possibilities: most of the countries concerned have only raw materials for export.
- Decreasing export prices: The world commodity indexes based on constant prices, with 1979–81 = 100, had decreased to 71 in 1988.[57] This development is expected to escalate owing to trade negotiations.[58]
- Energy supply: The USSR and China process nearly 70 per cent of the world's recoverable reserves of bituminous coal and 50 per cent of lignite and sub-bituminous coal.[59] The USSR also has nearly 40 per cent of the world's recoverable reserves of natural gas. It is difficult to see how these countries – under the economic conditions outlined above – will change to alternative energy resources.

Cost-Benefit Analysis

It must be stressed that to date, no complete cost-benefit analysis on climate change exists. This is probably because of the dilemma of global "commons" involving long-term decisions and effects as well as tremendous uncertainty. Another reason is that assessment tools and methods for addressing equity issues are badly underdeveloped.[60] When and where climate change will happen raises serious and fundamental problems. The best-developed body of relevant theory to deal with the trade-offs of climatic change is that:

> concerning the discounting of future costs and benefits to obtain single "preventative" estimates. There is little question that such discounting practices have a role to play where the question at hand is the comparison of alternative time flows of money or its equivalent. There is equally little question that the long lag times and potential irreversibilities associated with the greenhouse gas issue pose serious technical problems for the comprehensive use of discounting methods to weigh the future against present costs and benefits.[61]

The basic question is what are we going to bargain with future generations about. There is an urgent need to articulate the value-laden choices that underlie this bargain.[62]

Traditional economists hold that climate has little economic impact upon advanced industrial societies.[63] The effects of climate change in developed countries will be less than 1 per cent on

GNP over the next half century, they say. They agree that poorer countries with a large agricultural sector may be more vulnerable. In some "ballpark" cost estimates of a 2.5°C global warming, the economic impact is estimated at 3 per cent of the gross global product,[64] but for developing countries it could amount to an unaffordable 4 per cent of GNP because of a more vulnerable agriculture sector. In the calculations above, the values for biological diversity, amenity and environmental quality are not accounted for. Noone has analysed these values, argue traditional economists such as Nordhaus: therefore, we do not know whether the costs are low or high.[65]

An estimate of the damage costs of fossil fuel has been made from a climatic point of view; this came to US$3.6–36 $/ton CO_2.[66] This has been revised to US$10–36 $/ton CO_2.[67] IPCC (1990) has called for a 60 per cent reduction in the annual greenhouse emissions. If emission figures from 1985 (see Table 5.1) are used, that would mean a reduction of roughly 18.5 million tons of CO_2-equivalent greenhouse gases. Using the range of costs worked out by Nordhaus (1989), we obtain a yearly damage cost of US$61–666 million. This cost estimate does NOT include any environmental or amenity aspects. Nordhaus estimates that the IPCC call for a 60 per cent reduction in emissions would cost over US$300 billion annually, and argues that the first 10–20 per cent cut could be achieved fairly inexpensively but that further reductions would be extremely costly. (This is illustrated in Figure 5.1.) He argues that the possible reduction of emissions should be done efficiently and gradually, although judging from existing economic analysis, the best investment today is in learning more about climate change rather than in trying to prevent it. The calculations above are based on traditional economic approaches following the discounting principle.[67]

A number of partial cost-benefit analyses have been conducted. One takes the components of the climate change problem, comparing the economic cost of cutting carbon emissions by substituting coal-fired electricity with nuclear electricity, natural gas, and a variety of energy-efficiency measures and finding, in most cases, negative costs.[68] This means taking into account both cutting carbon emissions and, at the same time, saving consumers' money.

The cost of cutting emissions in the UK and Poland has been calculated.[69] The analysis shows that emissions could be cut by

Figure 5.1: Estimated annual worldwide cost of cutting greenhouse gases.*

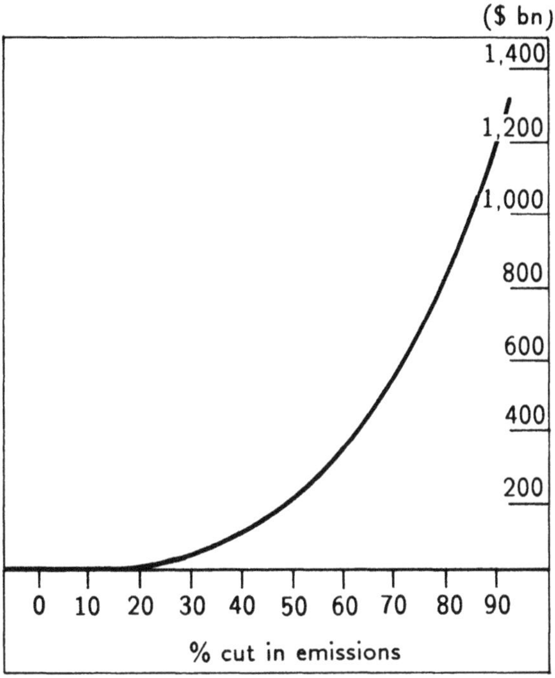

At 1990 levels of world economic activity. Assumes cuts through efficient policies phased in gradually.

Source:Nordhaus (1990)

20 per cent in the UK with substantial growth in the economy and rising living standards. For Poland it was found that the energy growth (66 per cent) at current practice for the period 1987–2005 could be cut to 22 per cent at a cost of US$30 billion. The cost of cutting CO_2 emissions to the required level have been worked out as US$15–20 *per capita* year in the USA, US$2–3 in Brazil and Jamaica, and much less for Asia, India and China.[70] The cost of taking no action against climate change, resulting in lower production in US agriculture because of less moisture, has been put at tens of billions of dollars.[71]

The US Agency for International Development argues that using present energy and technology prices as a basis for assessing long-term technological and policy alternatives – as is done in most traditional cost-benefit analyses – is misleading. It indicates that neither the present environmental costs of emissions nor the future costs of climate change are reflected in present prices of goods and services.

A first step towards economic evaluation of the tremendous long-term uncertainty involved in climate change has been produced by Yohe.[72] The basic problem lies in evaluating possible responses and identifying the appropriate timing for different decisions with only the present state of knowledge to work on. According to another school of economics, there is no such thing as cost of irreplaceable resources or of irreducible pollution.[73] This school also argues that no relevant price can be set on avoiding losses by emissions. The correct price can be set only by future generations' bidding. Unfortunately, future generations are unable to do that at present. Most of the historical depletion has taken place because "the price was right", according to traditional economics. Therefore, this school suggests that the only way out of the trap is quantitative restrictions.

In its policy declaration, *Ecology and the Global Economy*, the InterAction Council concluded that "current instruments are inadequately used to support the process of reconciling the competing environment and socioeconomic objectives. They provide policymakers, planners, and investors with insufficient, often wrong, signals, incentives and definitions of economic growth."[74]

At the moment, it is obvious that the existing cost-benefit analyses are of no help in formulating efficient countermeasures to climate change. That leads us to a policy approach following the principal of quantitative restrictions. Some research reports have taken up this concept and allocated a global budget for emissions,[75] divided equally between industrialized and developing countries. The national budget is based on *per capita* emissions. The basic concept is a time limit (number of years) for individual countries before they deplete their emission rights. Subak and Clark suggest approaches following CO_2 release patterns.[76]

Conclusions

The problems involved in formulating a global strategy for climate change are difficult. Feldman has defined the following:[77]

- identification of the atmosphere as a shared sustainable resource;
- agreements on causes, rates, effects and responsibilities;
- minimizing economic efforts while maximizing economic growth;
- encompassing the complexity in a multilateral policy.

We have discussed the major causes of greenhouse gas emission and the possibilities of mitigating the effects, and come to some rather bleak conclusions:

- The theoretical possibility of improving energy efficiency is great, but it will be difficult to achieve.
- Substantial use of alternative energy sources will come up against numerous socioeconomic and environmental obstacles.
- Natural gases cannot be regarded as a major substitute for fossil fuels.
- Nuclear power raises too many questions for it to be suitable either.
- The Montreal Protocol on CFCs needs strengthening, and it remains uncertain how this will be done.
- It is unrealistic to expect a rapid reduction in ongoing tropical deforestation.
- The possibility of decreasing agricultural emissions seems limited because of demographic factors.
- Large-scale forest plantations seem an unrealistic means of mitigating the greenhouse gases.
- No one seems able to deal with the problem of population. Georgescu-Roegen argues that the difficulty with population is human nature and a deep-rooted mistrust "of the rich that the poor will not cease growing in numbers and of the poor that the rich will not stop getting richer."[78]
- The pessimistic scenarios for socioeconomic development in the Third World and USSR/ Eastern Europe.

The expectations of world governments are probably too high. Sand says that the size of the problems caused by climate change means that they cannot be solved by world governments alone.[79] His view is supported by the Nordic UN Project, which concludes that there is "no common understanding, within or outside the

UN system, as to what the operational implications of sustainable energy development should be". The project also underlines the lack of consensus between developing and developed countries on the extent to which environmental issues should be administered by the UN system.[80]

Schelling is pessimistic about the possibility of tackling and slowing down global warming. He points out that actions from individual countries have only minor effect; therefore, individual countries will not act until international agreements on the reduction of greenhouse gases are enforced.[81] White foresees great difficulties in reaching an international agreement on mitigating climate change. He estimates that it will take a long time – even longer than the Law of the Sea Treaty, which took fifteen years.[82]

Managing greenhouse gases certainly depends more on political, cultural and socioeconomic conditions than on physical constraints.[83]

Notes

1. International Panel on Climate Change (1990); Chandler, W.U., *Carbon Emissions, Control Strategies* (Baltimore, MD: World Wildlife Foundation and The Conservation Foundation, 1990); Sinyak, Y., *Global Energy/CO_2 Projections*, Working Paper WP-90-51 (Laxenburg, Austria: International Institute for Applied Systems Analysis, 1990).
2. The scenario is based on work conducted by the US/Japan Expert Group (1990), the Task Force group of IPCC (1990) and scenarios developed by the EIS, AFOS and RSWG groups of the IPCC.
3. Goldenberg, J., Johansson, T.B., Reddy, A.K.N. and Williams, R.H., *Energy for a Sustainable World* (New Delhi: John Wiley & Sons, 1988).
4. Fischer, D. (ed.), *Options for Reducing Greenhouse Gas Emissions* (Stockholm Environment Institute, 1990).
5. Diepstraaten, F.M.J.A. and van Gool, W., "Prospects for carbon dioxide emission reduction", in Okken, P.A., Swart, R.J. and Zwerver, S. (eds), *Climate and Energy: The Feasibility of Controlling CO_2 Emissions* (Dordrecht: Kluwer Academic Publishers, 1989).
6. Becht, H.Y. and van Goest, J.P., "Energy conservation for a long-term sustainable energy policy", in Okken, Swart and Zwerver (eds), 1989.
7. Fulkerson *et al.* (1988); Mabro, R. and Davison, A., "The environmental impact of present and future energy consumption", Oxford

Institute for Energy Studies; presented at a high-level expert group on ecology and energy options, InterAction Council, Montreal, 29–30 April 1989; US Agency for International Development, *Greenhouse Gas Emissions and the Developing Countries: Strategic Options and the USAID Response: A Report to Congress.* (Washington, DC: July 1990); Ayres, R.U., *Energy Inefficiency in the US Economy: A New Case for Conservation*, Research Report RR-89-12 (Laxenburg, Austria: International Institute for Applied Systems Analysis, 1989); Ayres, R.U., Energy Conservation Seminar given at the International Institute for Applied Systems Analysis, February 1989; *Die Presse.* Genfer Klimakollaps. Konferenz ohne kondretes Ergebnis. *Die presse*, 7 November 1990; Lovins, A.B., "Energy, people and industrialization", paper commissioned for the Hoover Institution conference: "Human Demography and Natural Resources". The Hoover Institution, Stanford University, California, 1–3 February, 1989; Gibbons, J.H. and Blair, P.D., "Energy effiiciency: its potential and limits to the year 2000", in Helm, J.L. (ed.), *Energy, Production, Consumption and Consequences* (Washington DC: National Academy Press, 1990).

8. Lysen, E., "The potential of renewable energy to reduce CO_2 emissions", in Okken, Swart and Zwerver (eds), 1989.
9. Sedjo, R., "Forests: a tool to moderate global warming?", *Environment*, vol. 31, no. 1 (1989), pp. 14–20.
10. Goldenberg *et al* (1988).
11. Mabro and Davison (1989).
12. Cavanagh, R., Goldstein, D. and Watson, R., "One last chance for a national energy policy", in Abrahamson, D.E. (ed.), *The Challenge of Global Warming* (Washington, DC: Natural Resources Defence Council, Island Press, 1989).
13. Rotty, R.M. and Masters, C.D., "Carbon Dioxide From Fossil Fuel Combustion: Trends, Resources and Technological Implications", in *Atmospheric Carbon Dioxide and the Global Carbon Cycle*, Trabalka, J.R. (ed.), DOE/ER-0239, pp. 65–80 (Washington DC: Oak Ridge National Laboratory for Department of Energy, Carbon Dioxide Research Division, December 1985.
14. Victor, D.G., *Greenhouse Gas Emissions from High-Demand, Natural Gas-Intensive Energy Scenarios*, Working Paper WP-90-01 (International Institute for Applied Systems Analysis, 1990).
15. Blix, H., "Nuclear power and the environment", Paper presented at the Conference on Nuclear Power and the Changing Environment, British Nuclear Forum, London, 4 July 1989.
16. Keepin, B. and Kats, G., "Greenhouse warming: comparative analysis of nuclear and efficiency abatement strategies", *Energy Policy*, December 1988.
17. Ibid.
18. Mabro and Davison (1989).
19. See, for example, the International Hamburg Conference, 1989.
20. Sinyak (1990).

21. Shaw, R.W., *Global Security: The Threats are Changing*, Working Paper WP-90-56 (Laxenburg, Austria: International Institute for Applied Systems Analysis, 1990).
22. German Bundestag (eds), *Protecting the Earth's Atmosphere: An International Challenge* (Bonn: German Bundestag, 1989).
23. Grubb, M., *Energy Policies and the Greenhouse Effect* (Hampshire, UK: Dartmouth Publishing, 1990).
24. Lanly, J.P., *Tropical Forest Resources*, FAO Forestry Paper 30 (Rome: FAO, 1982); Mayers, N., *Deforestation Rates in Tropical Forests and their Climatic Implications* (London: Friends of the Earth, 1989); World Resources Institute (WRI), *World Resources 1990–1991, A Guide to the Global Environment* (Washington, DC: WRI, 1990); Food and Agricultural Organization (FAO), "Forest resources assessment 1990. Methodology for deforestation and forest degradation analysis", unpublished paper (Rome: FAO, 1990).
25. Palo, M. and Salmi, J., *Deforestation or Development in the Third World? Volume I*, Division of Social Economics of Forestry, Report 272, (Helsinki: The Finnish Forest Research Institute, 1987); Palo, M. and Salmi, J., *Deforestation or Development in the Third World? Volume II*, Division of Social Economics of Forestry, Report 309 (Helsinki: The Finnish Forest Research Institute, 1987); Palo, M. and Salmi, J., *Deforestation or Development in the Third World? Volume III*, Division of Social Economics of Forestry, Report 349, (Helsinki: The Finnish Forest Resource Institute, 1987).
26. Houghton, R.A., "The global effects of tropical deforestation", *Environmental Science and Technology*, 24 April 1990; Brown, L. and Flavin, C. in Brown, L. *et al.* (eds) *The State of the World* (New York: Norton 1990); Mayers, N. (1989).
27. Houghton, (1990).
28. Palo and Salmi (1987).
29. Nilsson, S., "Future development of tropical forest resources", unpublished manuscript (Laxenburg, Austria: International Institute for Applied Systems Analysis, September 1990).
30. Wiersum, K.F., "Reforestation, a feasible contribution to reducing the atmospheric carbon dioxide content?", in Okken, Swart and Zwerver (eds), 1989.
31. Freer-Smith, P.H., *Climate Change: The Contribution of Forests to Response Strategies*, Research Report 189. Forestry Commission Survey (1990).
32. Wiersum (1989), quoting an estimate by Meyers.
33. Postel, S. and Heise, L., *Reforesting the Earth*, Worldwatch Paper no. 83 (Washington, DC: Worldwatch, 1988).
34. Sedjo (1989).
35. Kohlmaier, G.H., Janecek, A., Benedroth, G., Kindermann, J. and Klaudius, A., "Land biota: source or sink of atmospheric CO_2. Positive and negative feedbacks within a changing climate and land use development", paper presented at the conference "Atmosphere, Climate and Man", Turin, January 1989.
36. Wiersum (1989).
37. Sedjo (1989).
38. WRI (1990).
39. German Bundestag (1989).
40. Sedjo (1989).

41. Nilsson, S., *Forest Resources, Forest Management and Diaper Fluff Pulp vs. Other Wood Uses*, (Procter & Gamble, FRG, 1991).
42. Wiersum (1989).
43. Sedjo (1989).
44. Ibid.
45. Nilsson, (1991).
46. Sedjo (1989).
47. Ibid.
48. US Agency for International Development (1990).
49. International Panel on Climate Change (1990).
50. Ciborowski, P., "Souces, sinks, trends and opportunities", in *The Challenge of Global Warming*. Abrahamson D.E. (ed.), 1989.
51. Ibid.
52. Beckman, G. and Klopries, B., "CO_2 – Austeig in der Troposphare: ein kardinal Problem der Menschheit", *Der Lichtbogen*, 208 (July 1989), pp. 4–31.
53. Revelle, R.R., "Thoughts on abatement and adaptation", in Rosenberg, N.J., Easterling III, W.E., Crosson, R.R. and Darmstadter, J. (eds), *Greenhouse Warming: Abatement and Adaptation*, RFF proceedings (Washington, DC: Resources for the Future, 1989).
54. Second World Climate Conference (1990).
55. Ayres (1989b).
56. WRI (1990); Lukina, I.V., "External debt of the USSR: history, trends, problems", *Russian Business Review*, launch issue (1990).
57. WRI (1990).
58. "Aftermath of the Rome summit", *Financial Times*, 30 October 1990.
59. WRI (1990).
60. Clark, W.C., *Usable Knowledge for Managing Global Climatic Change* (The Stockholm Environment Institute, 1990).
61. Ibid.
62. Koopmans, T., "Comments", pp. 322–3, in Lind *et al.*, *Discounting for Time and Risk in Energy Policy*, (Baltimore, MD: Johns Hopkins Press for Resources for the Future, 1982); quoted by Clark (1990).
63. Nordhaus, W.D., "Greenhouse economics – count before you leap", *The Economist* 7 July 1990.
64. Ettinger, J., Van Jansen, T.H. and Jepace, C.J., *Climate, Environment and Development*, Noordwijk Conference Report, Vol. I (1989).
65. Nordhaus (1990).
66. Ibid.
67. Ibid.
68. Chandler, W.U., "Global energy strategies and climate change", in *Proceedings, Second North American Conference on Preparing for Climate Change* (Washington, DC: Climate Institute 6–8 December 1988).
69. Environment Canada, "Cutting emissions of carbon dioxide: case studies of Poland and the United Kingdom", Energy Note 22

(Ottawa: Office of the Science Advisor, Environment Canada, 1989).

70. Goreau, T.J., "Balancing atmospheric carbon dioxide", *Ambio*, vol. 19, no. 5 (1990), pp. 231–6.
71. Adams, R.M., McCarl, B.A., Dudek, D.J. and Glyer, J.D., "Implications of global climate change for Western agriculture", *Western Journal for Agriculture Economics* 13 (1988), pp. 348–56.
72. Yohe, G.W., "Uncertainty, climate change and the economic value of information: an economic methodology for evaluating the timing and relative efficiency of alternative responses to climate change", in Clark, W.C. (ed.), 1990.
73. Georgescu-Roegen, N., "Energy and economic myths", *Southern Economic Journal*, vol. 41, no. 3 (1975), pp. 347–81.
74. InterAction Council, *Ecology and the Global Economy*, Report on the recommendations and conclusions of a high-level expert group (Amsterdam: InterAction Council, 10–11 February 1990).
75. International Project for Sustainable Energy Paths (1989).
76. Subak, S. and Clark, W.C., "Accounts for greenhouse gases: towards the design of fair assessments", in Clark W.C. (ed.), 1990.
77. Feldman, D.L., *Managing Global Climate Change through International Cooperation: Lessons from Prior Resource Management Efforts*. Prepared for Office of Environmental Analysis. US Department of Energy, by Global Environmental Studies Center (Oak Ridge, TN, July 1990).
78. Georgescu-Roegen, N. (1975).
79. Sand, P.H., *Lessons Learned in Global Environmental Governance* (Washington, DC: World Resources Institute, 1990).
80. The Nordic UN Project, *Towards Sustainable Energy Development, The Energy Activities of the UN System and the Development Banks* (Stockholm: The Nordic UN Project Report I, 1989).
81. Schelling, T., *Global Environmental Forces* (Harvard University: Environmental Policy Center, 1988).
82. White, R.M., "The great climatic debate", *Scientific American*, vol. 263, no. 1 (1990), pp. 18–25.
83. Grubb, M. (1990).

Chapter 6

International Responses

Despite the pessimism of some observers, there is at present a vigorous international response to climate change, involving research, discussion, planning and implementation even leading to – or being led by – a comprehensive legal convention. Two main agencies are involved: the United Nations Environmental Programme (UNEP) and the World Meteorological Organization (WMO). UNEP and WMO established the Intergovernmental Panel on Climate Change (IPCC) in 1988 after a resolution of the UN General Assembly. The IPCC was charged with assessing the scientific evidence on what was happening (Working Group 1) and the impact (Working Group 2) and responses (Working Group 3) to any changes. It was instructed to report to the Second World Climate Conference in Geneva in October 1990, from which a very general declaration was duly produced (Appendix II). The scientists were clearly worried, but the political documents did not advocate or facilitate commensurately dramatic action and the politically unimportant mountains were virtually missed out. In an attempt to reach consensus, the lowest common denominator prevailed. The Gulf crisis did not help. The shortage of oil, or the fear of shortage, constrained the chances of carbon restrictions in the leading industrial countries, where lifestyles are based on high energy consumption.

Most of the detailed recommendations concerning mountains were contained in the Report of Working Group 2, which is summarized in Appendix I. Although the focus of the group was on the cryosphere (defined as "seasonal snow cover, near-surface layers of permafrost and some masses of ice"), there was much reference and relevance to mountains. The Working Group 2 report was based on assumptions at the higher end of the climate-change estimates, despite the more guarded position of the WG1 report. Most of the scientific evidence was gathered before WG1 changed tack, and the calculations could not easily be redone.

The conclusion from WG2 was that the cryosphere would be "substantially" reduced by warming and that there would therefore be very significant impacts and socioeconomic effects. WG2's major recommendations for action related to the importance of getting better data and improving the models as well as assessing the needs, especially for protected areas, and providing educational materials (see Appendix I.) The scientific conclusions to the World Climate Conference stressed the need for more research, especially on the role of ice.

The research response is also prominent in the other international agencies working on climate change in the mountains. UNESCO's Man and the Biosphere programme, which had established a special project on mountains in 1973, had a long-standing interest in climatic interactions with the wider environment. The UN University Highland–Lowland Interaction Project worked within a context where mountains were considered in their interactions, especially with related river systems. FAO has worked on mountain watersheds.

Large sums of money are currently being committed to programmes which are duplicating each other or reinventing the wheel. Very few programmes are action orientated. Some action elements were included in International NGO's programmes, such as the IUCN, where protected areas were seen as a means of mitigating some of the effects of climate change. (See Appendix IV.)

The research emphasis is necessary but not sufficient. Action is needed before all the research is in, not least because an overwhelming proportion of the evidence points in the same direction and because it is clearly necessary to attack the root problem of the generation of greenhouse gases as soon as possible. But how can this action be implemented, especially at international levels? Working Group 3 of the IPCC produced a set of response scenarios (summarized in Appendix V), a shopping list of measures needed to stabilize and bring down emissions and concentrations. In the event, the World Climate Conference, in its final Ministerial Declaration (Appendix II), opted to produce – much to the annoyance of the NGOs – few of the targets suggested in WG3, offering instead some rather pious hopes. Greenpeace saw an unholy alliance between "climate criminals" – the USA, USSR, UK and Middle Eastern Oil States.

Many scientists were also disappointed. There was a very large measure of agreement amongst those working in WG1 concerning the probability of the greenhouse effect. The WG3 report was called "full of politics" by Bolin (Greenpeace 1990). But in fact political influences were obvious from the start in all the working groups. These influences were reflected in the chairmanship and, to some extent, the stance of the different working groups in the IPCC. The United Kingdom chaired Working Group 1 which gathered together the scientific evidence; the Soviet Union the group on the effects(2) and the USA that on responses(3). The groups had communication problems, as the *New Scientist* put it (23 June 1990), working mainly in parallel, though there should have been a linear sequence from WG1 to the others. The situation was further complicated when in mid 1990 WG1, for reasons that remain obscure, effectively shifted the goal posts, toning down the temperature estimates and estimates of sea-level rises.

If the responses at the end of the World Climate Conference really fudge the issues, what international approaches are necessary, particularly in the mountains? Three interrelated approaches may be identified: legal regulation; education; market forces. The last may be discounted to some extent. Some have argued that much of environmental economics represents market failure in that there is a basic contradiction between short-term profit-seeking and the long-term stability necessary for optimum environmental management. None of this excludes the operation of financial incentives and disincentives or price signals, nor of a fundamental role for consumer demand. Ultimately, it is the consumer who pays for environmental degradation.

Fiscal measures have been suggested – such as the "polluter pays" principle or emission trading – where companies (or countries) may pollute in one place above norms, provided that other sites are below. Neither has proved practical. The history of the "polluter pays" principle is not overly optimistic. It was once thought to be the panacea for all environmental degradation, but failure has now been generally admitted, mainly because the buck could be passed. Charges, fines, etc., can be converted into licences or passed on down the consumer chain. Taxes may be avoided or evaded. Carbon price rises do not necessarily work either. Automobile fuels are rather inelastic factors, and fuel accounts only for 10 per cent of the costs of running a vehicle. Anyway, as we have noted, carbon has a long life in the atmosphere.

Other schemes suggested have been to establish an insurance fund, increasing existing insurances by 1–2 per cent, or using the peace dividend released by demilitarization. Apart from the complications of such schemes and the bureaucracy so created, there is not much evidence that consumer and public behaviour would be altered to the extent necessary to achieve a reduction in pollution.

A major international effort at regulation has been to establish intergovernmental conventions for pollution control. In 1979 a Convention on Long Range Transboundary Air Pollution (CLRTAP) was concluded by the nations of the industralized countries through the United Nations Economic Commission for Europe (which also includes North America). In 1987 a protocol was signed requiring a 30 per cent reduction of SO_2 emissions and in 1989 a second for the phased reduction of NO_x and negotiations are taking place on VOCs. Some have rated CLRTAP a success, though the process has involved much slow diplomatic wrangling and there has been a good deal of backsliding. Certainly the number of trees dying in the mountain regions of Europe has continued to increase.

The most rapid international action has been directed towards the prevention of ozone depletion because of the deadly nature of ultraviolet radiation. A large number of countries met in Montreal in 1987 to reduce the use of CFCs and halons – the former by 20 per cent in 1994 (50 per cent by 1999); the latter to be eliminated three years after ratification of the agreement. At the London Conference in 1989 more stringent restrictions were proposed. The efficiency of ozone prevention, however, has been queried, as there are exceptions for countries which have lagged behind industrially. Moreover, there is at present a large stock of CFCs in existing refrigerators which will find their way into the atmosphere.

The restriction of CO_2 emissions has been a slower process. Tables 6.1 and 6.2 summarize the targets that have been adopted up to October 1990 in a number of leading countries, as well as desirable goals; more promises were made at the World Climate Conference in November, though at best these looked forward to stabilization by the year 2000. Although scientists have estimated that an immediate 50–80 per cent carbon cut is required, much less is politically acceptable – the best is probably a 20 per cent reduction by the year 2005, which has formed the basis for a

Table 6.1: Global carbon emissions 1988 and goals for 2000 and 2010

	1988		2000		2010	
Area	**Carbon (mio tons)**	**Per capita (tons)**	**Carbon (mio tons)**	**Per capita (tons)**	**Carbon (mio tons)**	**Per capita (tons)**
North America	1,379	5.07	897	3.03	662	2.13
Soviet Union and Eastern Europe	1,428	3.55	964	2.23	872	1.91
Oceania	336	2.27	284	1.79	270	1.65
Latin America	910	2.09	803	1.46	764	1.18
Western Europe	774	2.03	699	1.79	664	1.67
Middle East	187	1.14	187	0.83	217	0.74
Africa	534	0.86	646	0.73	749	0.64
Centrally Planned Asia	774	0.66	932	0.69	1,082	0.73
Far East Asia	833	0.55	998	0.52	1,158	0.52
World	7,319	1.42	6,435	1.03	6,438	0.93

Source: Leggett, J. (ed.), *Global Warming – The Greenpeace report* (Oxford: Oxford University Press, 1990).

Table 6.2: National carbon dioxide targets

West Germany	25% cut by 2005
Netherlands	Stabilize by 1994 and 3–5% cut by 2000. Then substantial cuts thereafter
Denmark	20% by 2000 and up to 50% by 2030
Italy	Stabilization by 2000 and possible 20% cut by 2005
France, Belgium, Luxembourg	Support a Community proposal to stabilize emissions at 1990 levels by 2000
UK	Stabilize at current levels by 2005. Wide range of measures in Patten Report (1990)
USA	Bills pending in Congress for 20% cut (Oregon already has legislation for 20% cut by 2005)
Austria	20% by 2005
Switzerland, Norway, Sweden	Freeze at current levels. Sweden plans CO_2 tax (1991)
New Zealand	20% by 2005
Canada	Stabilize CO_2 emissions at current levels by 2000
Japan	Stabilize at "lowest possible level" by 2000
Second World Climate Conference–European Community, Australia, Austria, Canada, Finland, Iceland, Norway, Sweden, Switzerland	Stabilize at 1990 levels

Source: WWF

number of countries' goals. More palatable may be a one per cent per annum reduction suggested by Canada at the Geneva Climate Conference. As the tables show, even dramatic cuts will barely reduce CO_2 levels.

Figure 6.1: Impact on atmospheric CO_2 concentrations of cuts

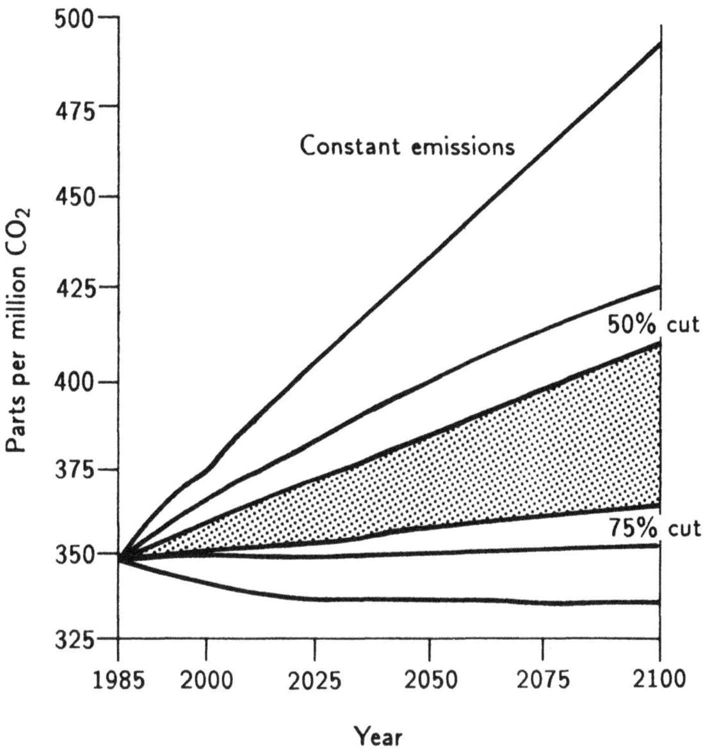

Source: Leggett, J. (ed.), *Global Warming – The Greenpeace Report* (Oxford: Oxford University Prss, 1990).

Current international action is directed towards a global convention (Appendix VI) that might attempt carbon reduction, as well as a protocol reversing deforestation. Such a convention is infinitely more difficult than the Montreal Protocol or the CLRTAP, since carbon control strikes at the heart of contemporary industrialization. In recent conferences such as the 1990

Ministerial Meeting on Sustainable Development convened by Mrs Brundtland in Bergen, leading industrial nations such as the USA and UK were careful to avoid precise targets and sent low-level delegations. The weak conclusion to the World Climate Conference merely continues this trend. The forestry proposals certainly have precise targets – for example 12 million hectares per annum to soak up the carbon – but some fear that this planting will be in inappropriate large-scale tropical forests, and not in the mountains.

To reduce carbon emissions does not require only legislation and incentives, especially in the industralized countries where most (70–90 per cent) of the emissions originate. Also required is a carrot-and-stick approach encouraging alternative fuels. There are clear figures showing the value of alternative energy sources (see Table 6.3).

Table 6.3: Costs of avoiding carbon emissions associated with alternatives to fossil fuels, 1989

Fossil fuel alternative	Generating cost (cents/kwh)	Carbon reduction (per cent)	Estimated pollution cost (cents/kwh)	Carbon avoidance cost ($/ton)
Improving energy efficiency	2.0–4.0	100	0.0	‹0.16
Wind power	6.4	100	0.0	95
Geothermal energy	5.8	99	1.0	125
Wood power	6.3	100	1.0	125
Steam injected gas turbine	4.8–6.3	61	0.5	97–178
Solar thermal (with gas)	7.9	84	0.2	180
Nuclear power	12.5	86	5.0	535
Photovoltaics	28.4	100	0.0	819
Combined-cycle coal	5.4	10	1.0	954

Source: Leggett, J. (ed.) *Global Warming – The Greenpeace Report* (Oxford: Oxford University Press, 1990).

The switch to alternative fuels has thus far been piecemeal, internationally. Nuclear accidents have made governments wary of this. There is sometimes a backlash effect. Sweden has said that having given up nuclear power, it cannot reduce fossil fuels. Although some emphasis is being given to solar hydrogen, natural gas, wind, geothermal and biomass, there is not much international action. Reforestation is still a minor activity without adequate international co-ordination. Only Europe and Japan are currently increasing their forested areas, despite the pressures of international NGOs devoted to this purpose, such as Global Releaf. In any case, reforestation must be accompanied by other measures which also reduce CO_2. Table 6.4 summarizes the possibilities. Mountains, where water, wind and insolation are to be found in abundance, could play a major role in this process, especially the production of solar hydrogen. In addition, there are vast reserves of non-polluting natural gas under the Alps.

Figure 6.2: Projected man-made CO_2 emissions (billion or 1,000 million tonnes carbon per year)

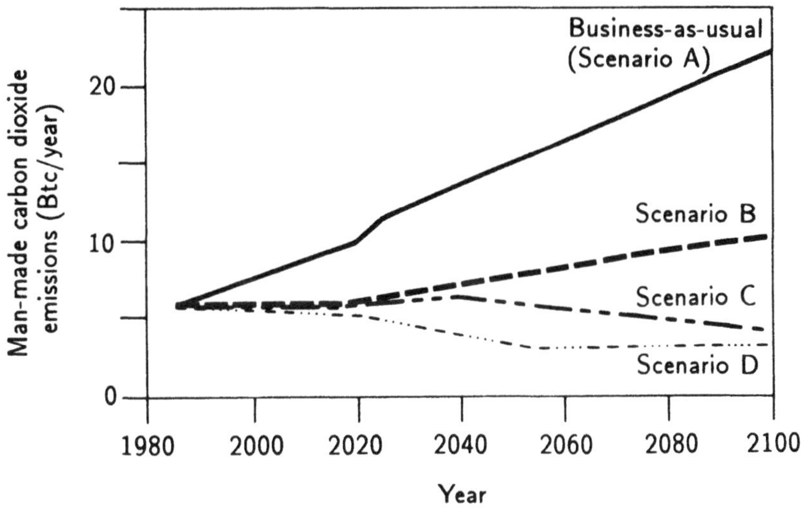

(See Appendix IV for explanations)

If legislation has not worked well, more may have been – and will be – achieved by improved education, training, awareness and communication. The key to carbon reduction is, after

Table 6.4: Potential CO_2 reductions from various response strategies given as gigatonnes of carbon per year – CTC/Yr

	1986	2005	2020
Historical emissions	6.8		
Projected emissions (base case)		8.5–9.5	10.0–12.3
Potential reductions from specific measures			
Efficiency		0.8–1.4	1.7–3.0
Shifts in fossil-fuel mix		0.2–0.2	0.2–0.3
Afforestation		1.5–1.5	1.7–2.0
Energy technologies			
Solar technologies		0.4–0.7	1.4–2.9
Wind		0.06–0.1	0.1–0.2
Geothermal		0.02–0.02	0.2–0.2
Biomass		0.4–0.4	1.0–1.0
Measures with severe environmental and/or sociopolitical limits			
Hydroelectric		0.0–0.1	0.0–0.5
Nuclear		0.0–0.3	0.0–0.6
Total of potential reductions		3.4–4.7	6.3–10.7
Feasible reductions, corrected for possible double-counting		3.0–4.5	5.2–8.8

Scale of potential reductions

	2005	2020
Feasible reductions as percentage of projected emissions	–30%--›–50% (~–40%)	–40%--›–90% (~–60%)
Difference between projected emissions levels (after feasible reductions) and the 1986 emissions expressed as a percentage	–50%--›–40% (~–25%)	+5%--›–80% (~–40%)

Source: Jager, J. (ed.), *Responding to Climate Change* (Stockholm: Stockholm Environment Institute, 1990).

all, changed behaviour. Certainly there have been efforts at comprehensive environmental education relating to mountains by UNESCO (1990) and UNEP (1990) involving the long-term reconstruction of school curricula as well as shorter-term goals of training key groups (decision-makers, professionals).

WWF, Mountain Wilderness and the Alpine Clubs have been active in raising awareness about mountain problems using a wide range of media. An interesting example of a recent campaign has been the Bellerive Foundations Alp Action project. This project, initiated by Prince Sadruddin Aga Khan in 1990[1], attempted to mobilize different sectors – science, conservation, government and industry – to publicize the plight of the mountains and successful grass-roots solutions (see Table 6.5). This book itself is a product of Bellerive's effort to promote more widely information about climate change and mountains. A business sponsor was identified for each project, to fund it and assist in advertising the messages within it. One of the great innovations of Alp Action is its pioneering efforts to convince and involve the business world. Such a task, which seemed impossible a few years ago, is now moving ahead. Alp Action has been able to utilize networks such as the World Economic Forum to alert the business community. A symposium at Davos in 1990 showed that there was a quite distinct green tinge to what was assumed to be a conservative world. Through Alp Action there are new joint projects between business and radical ecological NGOs.

But the NGOs do not have enough political clout to go it alone. What alternatives are there? At present the United Nations family do not even have enough teeth to chew over the necessary arrangements, let alone produce a legal digest. UNEP has only a small budget and is tied up in the red tape which plagues all UN bureaucracies. Plans do exist for radical changes. The (1989) Hague Meeting of seventeen leading industrialized countries proposed a new and strengthened institutional authority with powers to carry out a global warming agreement. Such a body would not only prepare for climate changes but, as in arms control, ensure that there were adequate verification measures. Such an authority must involve the blossoming NGO movement and the business world.

Any international authority requires a firm legal framework. How far have the negotiations for a Climate Convention gone in achieving this goal? In Appendix VI we reproduce the present

Table 6.5: Alp Action projects

1. THE RETURN OF THE BEARDED VULTURE: A living symbol for Switzerland's 700th Anniversary – Europe
2. RESTORING THAUR'S FIFTEENTH CENTURY PISCICULTURE: A guarantee for the future of endangered Alpine fish – Innsbruck, Austria
3. BERGWALD: Help conserve the protective functions of mountain forests – Switzerland
4. HIGH-QUALITY WATER SUPPLY FOR CHALETS IN THE PAYS D'ENHAUT ALPAGES – Switzerland
5. THE "SISTINE CHAPEL OF ITALIAN CERAMICS": Highlighting a seventeenth-century cultural heritage – Italy
6. ADOPT A BEAR: Don't be intimidated by the largest animal in the Alps – Austria
7. LOVELY LAC LAMOURA: Caring for a threatened ecosystems – France
8. TRANSALPINE EXPLORERS: A new adventure game on Alpine nature and culture – Europe
9. ALPINE GOAT-SHED AND CHEESE-DAIRY: Promote the production of "Greina Goat Cheese" – Switzerland
10. HEDGES MATTER: Raising children's awareness through field action – Liechtenstein
11. REVITALIZING THE INN RIVER'S ALLUVIAL FORESTS: a new enrichment for the Engadine landscape – Switzerland
12. BET ON THE LYNX: Help our largest cat recolonize the "triple border area" – Yugoslavia, Italy, Austria
13. WATCHING NATURE: A public observatory in a restored fourteenth-century watchtower – France
14. SOLAR PLANT FOR ELECTRIC CAR: Culture reborn in the Vanil Noir – Switzerland
15. THE FLIGHT OF THE MYTHICAL GRIFFON VULTURE: A summer dream comes true – Italy
16. PRESERVING AN ANCIENT HANDICRAFT: Culture reborn in the Vanil Noir – Switzerland
17. RESTORING A WETLAND PARADISE FOR FAUNA AND FLORA – France
18. CHILDREN'S SKI DISCOVERY OF ALPINE LIFE – Germany
19. FLOATING ISLAND: To shelter nesting seabirds in "Les Grangettes" – Switzerland
20. ALPINE DATABANK: A videodisk for the Château des Rubins – France
21. SOLAR ENERGY: A new power for schoolchildren – Switzerland
22. PRESERVATION OF THE INNSBRUCK PASQUE-FLOWER: Ensuring the survival of a highly threatened botanical marvel – Austria
23. A PROFESSIONAL HANDBOOK ON SOFT TOURISM HOTEL MANAGEMENT – Switzerland
24. CHILDREN PROTECT BIRDS IN THE YUGOSLAVIAN ALPS – Yugoslavia, Albania
25. SOLAR POWER PLANT: An initiation centre for renewable energy – France
26. WHAT FUTURE FOR ALPINE CULTURE AND LANDSCAPES?: A cinema trilogy will answer your questions – France

framework for discussion. In many ways, the analogy of the Montreal Convention and Vienna Protocols for Ozone are useful, since specific targets were involved. But any international agreements are only as good as the willingness of national governments to co-operate, as the experience of controlling the horizontal and vertical proliferation of nuclear weapons has shown. The experience of the International Court of Justice in The Hague has exposed the limitations of any international legal punitive measures. Monitoring and verification remain difficult – despite détente. The history of the Law of the Sea demonstrates how difficult it is for nations to relinquish their sovereignty. The cryosphere is not even defined in international law nor, apparently, to be covered in the proposed Climate Convention.

The response most needed and most lacking in the mountains is protection. Although it is not possible to erect a dome over the mountains and exclude air pollution or prevent climate change, there are still many benefits from protection. The system of protection established by the United Nations and administered by the World Conservation Union (IUCN) envisages different types of reserve, ranging from strict nature sanctuaries where humans are not permitted through limited protection in national parks to the biosphere reserve model where there are distinct areas for multiple uses and buffer zones. In the European region (apart from Iceland) no country has more than 15 per cent of their area listed as protected. The IUCN list (1985), however, covers only protected areas over 1,000 hectares. In countries such as Switzerland, there are many small village plots. In addition, there are lists of protected wetlands or cultural sites which are separately recorded.

Nevertheless, there is considerable scope for the extension of protection. Reinhart Messner (personal communication) and others have proposed that all permanent ice areas be made a wilderness. Most of the worlds ice is in Antarctica, and the present Franco–Australian initiative to establish a nature park there – and especially to restrict mineral exploitation – is an important step forward. Similar models have been suggested for the Arctic, the Himalayas, and elsewhere. The world's ice might well be administered by an international authority, possibly using some kind of trusteeship arrangement. Like the oceans, this would require an international Law of the Ice, which does not exist at the moment, and the elaboration of the ideas suggested by the Brundtland Commission and the First World Conservation Strategy (1980 and Dasmann and Poore)[2] to regard areas such as Antarctica as global commons. Even if such arrangements would not necessarily stop melting, at least wildlife and local cultures would be better protected.

Table 6.6: Protected areas in selected European countries and regions

Country	Protected area % Land area	ha/pop.
Austria	13.2	148.0
France	7.5	77.6
Switzerland	0.4	2.6
Italy	0.9	4.8
Sweden	2.3	125.6
Norway	1.5	126.9
Iceland	59.8	27,466.6
Czechoslovakia	7.9	67.2
Armenia SSR	6.8	69.3
Georgia SSR	2.0	28.0

Source: IUCN

There is certainly a need to think about innovative ways in which the cryosphere itself could be preserved. Here reforestation is important to prevent erosion, to provide a carbon sink, and because snow lingers longer, especially under conifer forest. The preservation of wetlands is also significant, for in many mountain areas the winters in anticyclonic conditions are cold enough but too dry. The case for extending wetlands or adapting existing dams to increase the frozen area in a controlled way is much more controversial. Snow cannons have adverse effects on flora and fauna. Cloud-feeding may also increase precipitation, though increases in turbidity require unacceptable smoke or pollution levels.

Most of these panaceas involve a technology where costs probably outweigh benefits. More significant are those responses which adapt to the changing climate, not only generic warming but also the sudden changes typical of present climate change or permafrost melting. Thus building and construction standards may be improved to withstand exceptionally heavy precipitation. Alternative tourist activities may be scheduled for winter resorts without snow – mountain biking, ice yachting, cultural events. In general, the rising tree line is extending the area for reforestation, but where dryness or acidity is a limiting factor, more efforts might be made to improve grassland management. Experiments from New Zealand, for example, have produced grasses which practically grow in lemon juice. Grass has very important functions in preventing erosion and retaining water. The tourist dimension should not be neglected either. Psychologists have pointed to the close bonding that people have to grassland landscapes, which may explain the current fetish for golf in mountain as in other regions.

Throughout there needs to be an emphasis on precautionary and preventive measures. An expanded campaign of education, training, awareness and communication can greatly assist this process. The increasingly common disaster situations require plans, equipment and an alert population. Funds, medicines, foods, etc., should be available but *in situ* in the mountains, where they are required. Most ecologists dislike military analogies, but some of the consequences of climate change are potentially as dangerous as nuclear war. The political and popular response must be to be forewarned and forearmed.

Conclusion

There is little point in reiterating the conclusions and recommendations of the IPCC, the Second World Climate Conference, the rash of new institutes climbing on the climate bandwagon, or even of the emerging mountain NGO pressure groups. Mountains are on the agenda for the 1992 UN Conference on Environment and Development (UNCED) and climate change will receive much attention. A network of mountain centres exists around the world; key NGOs, like the International Mountain Society and the Bellerive Foundation, are beginning to provide the necessary co-ordination and stimulus for this network; working closely with intergovernmental organizations.

All this is welcome and necessary, but not sufficient. Whatever is planned for the mountains to prepare for climate change can do little to turn the tide. Efforts to spray artificial snow on ski slopes can hardly delay the demise of winter tourism and will hardly change the albedo effect. The greenhouse gases and the pollutants will not go away for hundreds of years. Put crudely, the mountains are at the mercy of the changing elements, as they have always been.

What may be more readily – if not easily – changed are the social, economic and political structures that affect the mountains, at least in Europe. Robert Cox has claimed that the present is a watershed in modern history. For the first time in Europe since the Treaty of Westphalia (1648) the nation-state is threatened, both by regional integration and by divisive ethnic forces. Restructuring is in progress. An opportunity now exists for repairing the mountains – the roof of Europe – whilst the renovation of the European house is going on.[3]

A blueprint may be needed for mountain regions, promoting both conservation and sustainable development. In Europe, where there are ancient traditions and civilizations, there cannot easily be a white wilderness. None the less, most mountain landscapes and cultures could be protected. Such areas would have strictly controlled multiple uses (farming, tourism, conservation etc.) avoiding massive urbanization and industrialization. They might be car-free and would certainly be demilitarized. Despite détente, there remains a need for inviolable buffer zones, especially in times of danger and anarchy.

How protection can be extended over the mountains remains debatable. There is probably neither popular or political will to embark on a comprehensive programme for what are considered to be marginal lands. In a market-orientated economy – now in the East as well as the West – there is reluctance amongst consumers and businesspeople alike to abandon growth and high-energy pathways. Draconian measures, even if necessary, are unlikely from governments fearful of unpopularity and instability. Education and awareness, although hardly successful in the past, are a last hope.

Globally, we feel that a broad convention dealing with atmosphere, socioeconomic development, population, energy, technology and environment is required. The work carried out so far (see Appendix 6) is a first step. In our opinion, a gradual process of consensus, with all stakeholders taking part, is necessary. This can be achieved by a broad-scale convention. According to Nitze it is necessary to build this convention from a bottom–up process instead of the traditional top–down undertaking.[4] He recommends that the convention have a number of targets – for instance:

- the industrial countries should stabilize their emissions of greenhouse gases within *x* years of signing the convention;
- every signatory should agree on an energy efficiency of *y* percent within *x* years;
- every signatory would ensure no net loss of forests after *x* years, etc.[5]

Based on our earlier discussion of the possibilities of mitigating the greenhouse gases, we believe that Speth has, in an excellent manner, formulated the topics which must be covered in the proposed global convention. In his view, these are the required steps or topics:

- a demographic transition towards stable populations, both in nations where growth is explosive and on a global basis before the world's population doubles again;
- a technological transition away from today's resource-intensive, pollution-prone technologies to a new generation of environmentally benign ones;
- an economic transition to a world economy based on reliance on nature's "income", not on depletion of its "capital";
- a social transition to a more equitable sharing of environmental and economic benefits;
- a transition in information and consciousness to a far more profound understanding of global sustainability;
- an institutional transition to new arrangements among government and peoples that can achieve environmental security.[6]

The wheel should not be reinvented in this process. Apart from a climate-change convention, there are proposals for conventions on biological diversity and an international instrument on forests (Maini 1991). There are also existing international conventions which have relevance – including the conventions on Wetlands of International Importance the Protection of the World Cultural and Natural Heritage (Paris, 1972), the International Trade in Endangered Species (CITES) (Washington, 1973); not to mention the pollution instruments, e.g. ozone (Vienna, 1985; Montreal, 1987), transboundary air pollution (Geneva, 1979) and as regional documents designed to protect natural resources (Washington, 1940; Algiers, 1968; Brasilia, 1978; Bern, 1979; Kuala Lumpur, 1985).

The forest instrument in particular has great potential. According to a Report of the Secretary-General for the Preparatory

Committee of the UNCED (A/Conf 151/PC/27, 5 February 1991, p. 31) this instrument should:

- fully integrate all aspects of environmental protection and economic development as they relate to all forests, whether boreal, temperate or tropical;
- ensure a vital increase in international co-operation for policies and programmes aimed at forest conservation and sustainable use;
- complement other international initiatives with implications for aspects of forestry, notably the climate change and biodiversity conventions;
- ensure the incorporation of the concept of sustainable management of forests and its relation with food, fuel, shelter and timber;
- provide a framework for increased national efforts and bring together both the ecological and economic approaches to the use of forest resources;
- provide a mechanism to monitor its implementation, to assess progress and to co-ordinate international action.

So far, mountains have not been particularly prominent in the forestry discussions, and the case for their protection should be made here and in the other conventions we have mentioned.

A case may be made for a separate international convention for the mountains. There is little in international law on the cryosphere apart from the Antarctic treaty system, and ice is poorly defined, if defined at all. The objectives need not be only, or mainly, a set of mandatory prescriptions but rather "softer" legal instruments embracing guidelines, codes of ethics, etc. There has already been a series of international meetings which have set out what is known and what should be done, including Littlehammer (UNESCO, 1973), Mohonk and Honolulu (1986, 1988). UN University and International Mountain Society, Tshakadzor USSR (1989), (Geneva – Bellerive – 1989), Davos (Bellerive, 1990, 1991). The 1989 Geneva Meeting produced the "Bellerive Initiative" which, although concerned with the Alps, set out ten action steps which might be applied in any mountain region (see Appendix VII).[7]

Since every mountain range is different, a complementary approach may be to generate conventions for the twenty or more main ranges of the world. One model is the Alpine Convention, which has attempted, not altogether successfully, to harmonize several existing legal systems. Since many mountain ranges are international – indeed, are boundaries between countries and cultures – this task will involve delicate negotiations. An alternative is to elaborate microscale frames for valleys; this also has the advantage of incorporating highland–lowland interactions associated with the river systems and watersheds. These frames are not necessarily mutually exclusive; international conventions may serve as templates into which regional or local plans and frames may be slotted.

There is an urgent need for action, both because of the problems and because there are at present opportunities which may not reoccur. The UNCED 1992 Conference (usually called ECO 1992) in Rio de Janeiro is a unique event. Not since the Stockholm Conference of 1972 has there been such a major gathering of those who can influence the future of environment and development. The NGOs, renamed the independent sector, are being given a place at the table, even if they are not carving the joint. The independent sectors include the CONGO group (NGOs in consultative status with the UN Economic and Social Council), all other environment and development NGOs, business and industry, trade unions, professional associations, scientific and academic institutions, women's organizations, youth groups, religious and spiritual groups, indigenous peoples' organizations and other citizen groups.

Before Rio the mountain constituency must get its act together – not just to be on the agenda, but to push a pragmatic plan of action designed to develop as well as protect. There are suggestions for the latter (see Appendix III), even if they are disputed, but much less that is innovative for the former in keeping with essentials of sustainable development which will be the cornerstone of ECO 1992. Sustainable development has been defined as "development that meets the needs of the present without compromising the ability of future generations to meet their own needs" (World Commission, 1987). Implied in this concept is that growth is not quantitatively limited (the idea that dominated in Stockholm in 1972) but, through technology and human actions, can be qualitatively reorientated. In this process there must be an alliance of the best positive ideas, rather than conservative charities and bureaucracies. This means bringing in those environmentally conscious and responsible elements of the business community to help reformulate regulations, provide capital (in all its senses), and above all assist in communication, education, training, awareness and advertising environmentally benign products and services.[8]

What can the private sector do specifically? There have been some concrete suggestions.[2] Kramer, Chairman of CIBA Geigy (IEB, 1989), has argued that, first of all, the private sector must acknowledge that there is a problem and not fall into the trap of advocating no action just because there is speculation and uncertainty. Secondly, there must be a recognition that industry is part of the problem rather than shifting the blame to others with whom co-operation should be sought. Thirdly, the private sector should pool its technical, scientific and financial resources. This process has already begun through the International Chamber of Commerce, the International Environment Bureau or the group assembled for the World Conference on Environment and Development by Schmidtheiny, called the "green titans".

Corbett (Monsanto – IEB, 1989) has continued this line of thought by emphasizing that companies should do more than comply with the legislation; rather, they should take initiative before legislation. He cited examples of the voluntary phasing out of CFCs (Dupont), reducing hazardous emissions (Monsanto), waste elimination (Dow, Union Carbide, 3M) as well as widespread efforts for improved efficiency and energy conservation. Corbett also suggested that biotechnology might produce tree species more resistant to pollution and better able to cope with climate warming, though critics have seen this as dangerous "appeasement" of the pollution threat.

The private sector has also produced codes of ethics, such as the Valdez Declaration prepared after the Alaskan disaster. Alp Action (Newsletter No. 1, 1990) has featured this as a model for the mountains. Alp Action, a programme encouraging the corporate funding of environmental protection, has itself consistently emphasized the need for the recognition of a responsible attitude by industry and has awarded prizes to the environmentally conscious.

Finally, the private sector has shown itself willing to work with other NGOs or the independent sector, as it is now called. At the Bergen Conference convened by Mrs Brundtland, a mechanism was established for continuing contacts. Although critics have claimed that industry is merely window-dressing to catch the green consumer, the process of dialogue and negotiation must be engaged. The question is: how? The problem is that the constituencies in the NGO network have not got their act together. Although the 1990 Bergen Conference identified and brought together at least five of the NGO sectors (industry, trade unions, youth, environment groups, scientists), none focused clearly on the mountains. Since Bergen, as part of the Rio process, the scientists have attempted a "mountain agenda", organized by the International Mountain Society, the United Nations University and the International Centre for Integrated Mountain Development in Kathmandu. This initiative relies on individuals filling out an elaborate questionnaire which will be presented to governments at Rio, as a status report and manifesto.

The scientists' initiative is necessary but hardly sufficient. The International Mountain Society numbers scarcely a thousand members, very few resident in the highlands of the world where 500 million people live. The questions asked by the scientist are neither new nor always readily understandable at the grass roots. Not all scientists are disinterested in a free-market world where there is increased competition for grants and power.

There has been a tendency to skirt politically sensitive mountains (e.g. Antarctica) and to neglect the cryosphere, the preserve of the glaciologists. Antarctica, where 98 per cent of the world's ice is located, must be closely integrated with climate-change reforms because the continent affects the global climate in so many ways. In the past, the cosy club of Antaractic Treaty powers has excluded governments and NGOs alike. Even the Secretary-General of the United Nations is not invited to the consultative meetings. Antarctica, like the sea ice, is not national territory in international law, even if some nations claim it – something like a global common. The current ideas to create a protected world park in Antarctica may be a useful model for the rest of the cryosphere, if not the mountains. Certainly, a role should be explored for the UN, even if of late it has become the cat's paw of national interests. Very useful are the recent attempts by GIPRI, Bellerive and others to mediate between conflicting positions and to open up the debate to a wide public.[9]

For the mountains, an interesting alternative proposal has been put forward by the Bellerive Alp Action programme. Alp Action has consciously attempted to forge a fair and frank partnership between all the actors interested in mountains and to unite them in microprojects in which there is a strong participation, even initiation, from below. Alp Action has identified the key NGOs working in the field and has sought the widest possible dissemination of good, innovative ideas. It is working closely with the United Nations Conference on Environment and Development, but also with those who cannot easily fit into the bureaucratic and diplomatic niceties that surround all UN activities.

A series of next steps should be urgently considered, including:

- the fuller incorporation of mountains and the cryosphere in the World Climate Convention process. The cryosphere may want separate treatment in a Law of the Ice process;
- the immediate expansion of protected areas in specially designated mountain and mountain forest areas. A map for the Alps was prepared as long ago as 1974 for the Trent Conference (Club 1974);
- an intensive campaign of education, training and awareness;
- a coalition and network of interested NGOs and official bodies to facilitate the exchange of information and improve co-ordination. There are still major obstacles to communication between the cultures and language groups in Europe;
- increased participation at local level in reforestation schemes such as the Bellerive Alp Action Green Roof project or Global Releaf;
- more emphasis on the preservation of biodiversity through more comprehensive red data books and rescue operations;
- the preparation of templates for sustainable development in different mountain regions to feature soft tourism "distance" employment[10], an agricultural renaissance and community development;
- contingency plans and funds for surprises and disaster relief.

All this means that most ancient and difficult of tasks – changing human nature rather than just nature. There is a need to be much more aware of the cultural mosaic which has survived, especially in the mountains. Those who demand the changes come generally from the plains and have too little appreciation of mountain traditions. There will be a limited effect if there is no communications and co-operation with and sympathetic comprehension of the grass roots that struggle to survive in the harsh mountain environment.

Notes

1. Bellerive Foundation, *Towards Alpine Action* (Geneva: Bellerive Foundation, 1989; Geneva: Alp Action, 1990).
2. Dasmann, R. and Poore, D. *Ecological Guidelines for Balanced Land Use, Conservation and Development in High Mountains* (Gland: IUCN, 1979).
3. Cox, R., *Globalization, Multilateralism and Social Choice*, New York: United Nations University, Work in Progress, 13–1. 1990.
4. Nitze, W.A., "A proposed structure for an international convention on climate change", *Science*, 249 (1990), pp. 607–8.
5. Ibid.
6. Speth, J.G., Environmental Security for the 1990s. . .in six not-so-easy steps. Issues and Ideas. (Washington: World Resources Institute, January, 1990).
7. Pitt, D. and Stone, P. *Towards Alpine Action* (Geneva: Bellerive, 1989).
8. Schimdtheiny, S., "Environment and development", paper presented to the Academy of the Environment, University of Geneva (1990).
9. GIPRI, Antarctica - The Villa Rigot Reflection (Geneva: GIPRI, 1991).
10. International Labour Office (ILO), "1990 – working at a distance", *ILO Information*, vol. 26, no. 5 (1990).

APPENDICES

Appendix I

Impacts of Climate Change on Seasonal Snow Cover, Ice and Permafrost and Socioeconomic Consequences – the IPCC Conclusions

Major findings

- The global areal extent and volume of the terrestrial cryosphere (seasonal snow cover, near-surface layers of permafrost and some masses of ice) will be substantially reduced. These reductions, when reflected regionally, could have significant impacts on related ecosystems and social and economic activities.
- Thawing and reduction in the areal extent of the terrestrial cryosphere can enhance global warming (positive feedback on climate warming) through changes in the global and local radiation and heat balances, and the release of greenhouse gases. This positive feedback could increase the rate of global warming and, in some regions, result in changes that are sudden rather than gradual. The possibility of relatively rapid changes increases the potential significance of the associated impacts.
- The areal coverage of seasonal snow and its duration are projected to decrease in most regions, particularly at mid-latitudes, with some regions at high latitudes in the Arctic and Antarctic possibly experiencing increases in seasonal snow cover.
- Decreases in seasonal snow cover can have both positive and negative socioeconomic consequences owing to impacts on regional water resources, winter transportation and winter recreation.
- Globally, the ice contained in glaciers and ice sheets is projected to decrease. Regional responses, however, are complicated by the effect of increased snowfall in some

areas which could lead to accumulation of ice. Glacial recession will have significant implications for local and regional water resources and thus impact on water availability and on hydroelectric power potential. Enhanced melt rates of glaciers may initially increase the flow of meltwaters; however, flows will decrease and eventually be lost as glacial ice mass decreases. Glacial recession and loss of ice from ice sheets will also contribute to sea-level rise.

- Degradation of permafrost is expected, with an increase in the thickness of the seasonal freeze–thaw (active) layer and a recession of permafrost to higher latitudes and higher altitudes. The thickness of the active layer is expected to increase by 1 metre over the next 45–50 years. Although major shifts are expected in climatic zones, recession of permafrost will significantly lag behind, receding only 25–50 km during the next 40–50 years. These changes could lead to increases in terrain instability, erosion and landslides in those areas which are currently underlaid by permafrost.
- The socioeconomic consequences of these changes in permafrost could be significant. Ecosystems underlaid by permafrost could be substantially altered owing to terrain disturbances and changes in the availability of water. The integrity of existing and planned structures and associated facilities and infrastructure could be reduced by changes in the underlying permafrost. Retrofitting or redesigning would be required at a minimum; however, in some situations the associated terrain disruptions and/or costs (environmental, social and economic) may be too large, necessitating abandonment. Development opportunities could also be affected in areas where the risks associated with developing in an area susceptible to permafrost degradation are assessed as too high.
- The terrestrial cryosphere, because of its relative responsiveness to climate and climatic changes, provides an effective means of monitoring and detecting climatic change.
- Lack of sufficient data and gaps in the understanding of associated processes limit more quantitative assessments at this time.

Principal issues

The terrestrial component of the cryosphere consists of seasonal snow cover, mountain glaciers, ice sheets and frozen ground, including permafrost and seasonally frozen ground. These elements of the terrestrial cryosphere currently cover approximately 41 million square kilometres, with seasonal snow cover covering as much as 62 per cent of the Eurasian continent and virtually all of North America north of 35° latitude.

Projected changes in climate will dramatically reduce the areal extent and volume of these elements of the terrestrial cryosphere. This has implications not only with respect to changes in the availability of fresh water and changes in sea-level and terrain characteristics, but also for societies and related economic systems which have come to depend on – or are limited by – the existence of a terrestrial cryosphere.

Feedback mechanisms are an important factor in understanding the impacts of climatic change on the terrestrial cryosphere. Reduced areal coverage of these elements and degradation of permafrost as a result of climatic warming can enhance warming through changes in surface characteristics and release of greenhouse gases. The impacts of socioeconomic consequences of changes in the terrestrial cryosphere will depend to a large extent on the rate at which the changes occur. Where the rate of change is quick or sudden, environment and associated social and economic systems will have little time to adapt. Under these circumstances, the impacts and socioeconomic consequences could be considerable.

Seasonal snow cover

General Circulation Models indicate that in most parts of the northern and southern hemispheres the area of snow cover is expected to decrease as a result of increased temperature and, in most regions, a corresponding decrease in total mass of the snow. Areas where snow cover is projected to increase include latitudes south of 60°S and higher elevations of inland Greenland and Antarctica (though the latter is, and will remain, largely a cold desert).

A reduction in the areal snow coverage and in the length of

the snow-cover season will result in a positive climatic feedback, increasing global warming as a result of the greater amount of solar radiation that a snow-free surface can absorb relative to one that is snow-covered. Loss of snow cover has both negative and positive socioeconomic consequences. Decreases in snow cover will result in increased risks of damages and losses for those systems which rely on snow as protection (i.e. insulation) from cold winter climates. Included are agricultural crops such as winter wheat, trees and shrubs, hibernating animals, and construction and maintenance of municipal infrastructures.

Reductions in both the temporal and spatial extent of seasonal snow cover will have significant ramifications for water resources, as the amount of water available for consumptive (e.g. potable and irrigation water) and non-consumptive (e.g. hydroelectric power and waste-management) uses decreases. Particularly sensitive are those areas such as the Alps and Carpathians, the Altai mountains of Central Asia, the Syr Darya and Amu Darya region of the USSR, the Rocky Mountains and the North American Great Plains, all of which are dependent on snowmelt for the majority of their spring and summer water resources.

Changes in snow cover will also affect tourism and recreation-based industries and societies, particularly winter recreation sports such as skiing. Projected climate change could eliminate a $50 million per annum ski industry in Ontario, Canada.

From a positive-impacts perspective, reductions of seasonal snow cover will reduce expenditure on snow removal, increase access opportunities and ease transportation problems. A reduction in snow cover, however, will also have adverse impacts for transportation in those areas which rely on snow roads in winter. Inability to use snow roads will mean that other, more costly methods of transportation will have to be used.

Ice sheets and glaciers

The relationships between climate and ice sheets and glaciers are complex and, because of relatively limited monitoring and research, not fully understood at this time. Increased temperatures generally result in increased ablation, and hence in a decrease in ice mass. Conversely, increased snowfall usually increases ice mass. Since projected changes in climate for some ice-covered regions include increases in both temperature and snowfall, an understanding of the impact of climate changes on glaciers and ice sheets must involve a consideration of the combined impact.

The bulk of the earth's ice mass is stored in the Antarctic ice sheet, divided between an eastern portion resting on continental crust and a large western portion which is underlaid by both continental crust and ocean. Much of the remaining ice mass is contained in the Greenland ice sheet, with smaller quantities stored in glaciers throughout the world. Although observed data are limited, it is estimated that both the Antarctic and Greenland ice sheets are at present roughly in equilibrium, with annual gains close to annual losses. There is some evidence to suggest that the Greenland ice sheet has been thickening since the late 1970s; this has been attributed to the new snow accumulations on the ice sheet.

Greenhouse-gas-induced climatic change will tend gradually to warm these sheets and bring them out of balance with the new climate regime. Change in ice-sheet volume, however, is likely to be slow, with significant loss unlikely to occur until after 2100. Calculations for Greenland suggest that a 3 per cent loss of ice volume in the next 250 years is possible, based on the projected changes in climate. In the case of the Antarctic ice sheet, the situation is more complex. The mass of the eastern ice sheet is expected to remain virtually the same or increase slowly as a result of expected increases in precipitation and temperatures. In contrast, the western ice sheet, like other marine ice sheets, is inherently unstable. Climatic warming could cause groundline retreat and rapid dispersal of ice into the surrounding ocean by way of relatively fast-flowing ice streams. These changes in behaviour could lead to collapse of a portion of the western Antarctic ice sheet; this, depending on the amount of ice involved, could have a dramatic impact on sea level and the surrounding environment.

The response of glaciers to climatic change will depend on their type and geographic location. In general, however, they have been shrinking for the last hundred years and are expected to continue to do so in response to projected changes in climate. In Austria a 3°C warming by 2050 is projected to cause a reduction by about one-half in the extent of Alpine glaciers. Melting of glaciers in the Soviet Arctic archipelagos may result in their disappearance in 150–250 years. In contrast, an assessment of mountain glaciers in the temperate zone of Eurasia indicates that up to 2020 these glaciers will, in general, remain essentially unchanged, with increased precipitation compensating for increased melt.

Ice-sheet and glacier melting will result in higher sea levels. Observations over the last century indicate that levels have been rising between 1 an 3 millimetres a year, primarily as a result of mass loss from Alpine glaciers. Current projections suggest an accelerated rise with greenhouse gas warming to a most probable rise of 65 centimetres by the end of the next century. Glacial melting can act as a negative feedback to regional and global warming, with heat extracted from the air to melt glacial ice and snow, thereby reducing the degree of warming.

The melting of glaciers will also alter regional hydrologic cycles. In New Zealand it has been estimated that a 3°C increase in temperature would, in the short term, increase glacier-fed river flow in some western rivers, increasing hydroelectric power generation by 10 per cent. Another effect of glacier retreat is possible increased debris flows. Large amounts of debris masses on steep slopes will become exposed as a result of glacial retreat and, therefore, would be unstable and vulnerable to the effects of erosion. Landslides would result, leading to burial of structures, traffic routes and vegetation. Obstructions of river flows and increased sediment loads resulting in changes in water quantity (e.g. local floods and reduced flows downstream) and water quality would probably also occur as a result of debris flows.

Permafrost

Permafrost is the part of the terrestrial cryosphere consisting of ground (soil and rock) that remains at or below freezing throughout the year. It usually contains ice which can take a variety of forms, from ice held in soil pores to massive bodies of more or less pure ice many metres thick. The presence of this ice in the ground makes it behave uniquely as an earth material and makes its properties vulnerable to climatic warming. At present about 20–25 per cent of the earth's land surface contains permafrost, primarily in the polar regions but also in the alpine areas at lower latitudes. It occupies approximately 10.7 million square kilometres in the USSR, 5 million square kilometres in Canada, 2 million square kilometres in China and 1.5 million square kilometres in Alaska. Present and past climate is the major determinant of permafrost occurrence and characteristics; however, a variety of other factors are also important – for example, the properties of the soil, and overlying terrain vegetation and snow cover.

Permafrost is usually present where the mean annual air temperature is less than −1°C. At temperatures near this value it is discontinuous in extent (discontinuous permafrost zone). Both its extent and its thickness increase at progressively higher latitudes where temperatures are lower. It has been found to extend to depths of approximately 1,000 metres or more in parts of Canada, approximately 1,500 metres in the USSR and 100–250 metres in China.

Permafrost can also exist in seabeds. There is extensive ice-bound material in the continental shelf beneath the Arctic Ocean; however, this permafrost is commonly relict (i.e. it formed under past conditions and would not form under current ones).

Permafrost is to a large extent inherently unstable, since it exists so close to its melting point. Most responsive to changes in climate would be those portions nearest the surface. Climate warming would thicken the active layer, leading to a decrease in soil stability. This permafrost degradation would lead to thaw settlement of the surface (thermokarst), ponding of surface water, slope failures (landslides) and increased soil creep. This terrain instability would result in major concerns for the integrity and stability of roads, pipelines, airfields, dams, reservoirs and other facilities in areas which contain permafrost. Terrain instability of the surface layer can also occur as a result of permafrost degradation in alpine areas, such as the Alps. This instability could result in dangerous debris falls from thawed rocks and mudflows.

Slope failures, thermokarst and loss of near-surface moisture, as the increased depth of the active layer moved limited water supplies further from the surface, would have detrimental effects on vegetation and could lead to significant decreases in plant populations. In the longer term, permafrost degradation would allow the growth of deeper-rooted, broadleaved species and the establishment of denser forest of coniferous species. Wildlife could also be affected through changes in terrain, surface hydrology and food availability. Loss of species and habitats could be expected, especially where wetlands dry out or areas are flooded as a result of melt.

Assessment of the effects of climate change on permafrost in any particular location must consider factors other than temperature – for example changes in summer rainfall and snow cover. In general, however, the projected warming during the next several decades would significantly deepen the active layer and initiate a northward retreat of permafrost. It is expected that a 2°C global warming would shift the southern boundary of the climatic zone currently associated with permafrost over most of Siberia north and northeast by at least 500–700 kilometres. The southern extent of permafrost will lag behind this, moving only 25–50 kilometres in the next 40–50 years (up to 10 per cent reduction in an area underlaid by continuous permafrost). The depth of the active layer is expected to increase by 1 metre during the next 45–50 years. Projected changes in permafrost in Canada are of a similar magnitude.

The melting of permafrost would result in the release of methane and, to a lesser extent, CO_2 from previously frozen biological material and from gas hydrates. The extent to which this will enhance the greenhouse effect is uncertain, but could be about 1°C by the middle of the next century.

The socioeconomic impacts of permafrost degradation will be mixed. Maintenance costs of existing northern facilities such as buildings, roads and pipelines will tend to rise, with abandonment and relocation needed in some cases. Change in current construction practices will be necessary, as may be changes in sanitary waste disposal. Benefits from climate warming and permafrost melt are likely for agriculture, forestry, hunting and trapping.

Recommendations for action

Projected greenhouse-gas-induced changes in climate will lead to ablation of global ice masses. There is uncertainty, however, regarding how this global response will be reflected at the regional/ local level and how the individual ice masses and seasonal ice and snow will respond. The most important effects of climatic change at high latitudes and elevated regions will be on and through changes in the terrestrial cryosphere. Furthermore, the terrestrial cryosphere is particularly suited for early detection of the effects of climate change. These two points necessitate a better understanding of the nature and dynamics of these ice masses and the factors that control them. This will require:

- establishment or enhancement of integrated, systematic observation programmes commensurate with research on the use of more efficient ground-based systems and remote-sensing technologies designed to provide baseline information and trends;
- concurrent monitoring of those facilities, structures and natural resources that are at risk owing to projected changes in the terrestrial cryosphere;
- establishment of new guidelines and procedures for design and construction practices that consider the impacts of climatic changes on permafrost;
- research, including international co-operative efforts, on the relationships between components of the terrestrial cryosphere and climate in conjunction with other determining factors, including feedback mechanisms;
- refinement of existing climate-terrestrial cryosphere models;
- national and regional impact assessments that will provide data and information on the impacts of climate change on areas in which components of the terrestrial cryosphere occur, and the socioeconomic consequences;
- assessment of the needs for protected areas (natural reserves) for affected species and habitats; and
- development and distribution of relevant educational material and information on climatic changes, their impacts on the terrestrial cryosphere and socioeconomic consequences, as well as a wider distribution of research results.

Appendix II

Ministerial Declaration of the Second World Climate Conference (7 November 1990)

Preamble

1. We, the Ministers and other representatives from 137 countries and from the European Communities, meeting in Geneva from 6 to 7 November 1990 at the Second World Climate Conference, declare as follows:
2. We *note* that while climate has varied in the past and there is still a large degree of scientific uncertainty, the rate of climate change predicted by the Intergovernmental Panel on Climate Change (IPCC) to occur over the next century is unprecedented. This is due mainly to the continuing accumulation of greenhouse gases, resulting from a host of human activities since the Industrial Revolution, hitherto particularly in developed countries. The potential impact of such climate change could pose an environmental threat of an up to now unknown magnitude; and could jeopardize the social and economic development of some areas. It could even threaten survival in some small island States and in low-lying coastal, arid and semi-arid areas.
3. We *appreciate* the work of the World Climate Programme (WCP) during the past decade which has improved understanding of the causes, processes and effects of climate and climate change. We also *congratulate* the IPCC, established by the United Nations Environment Programme (UNEP) and the World Meteorological Organization (WMO) on its First Assessment Report on Climate Change. It has identified causes and possible effects and strategies to limit and adapt to climate change and, in the light of the United

Nations General Assembly resolutions, has identified possible elements for inclusion in a framework convention on climate change.

4. Recognizing climate change as a common concern of mankind, we commit ourselves and intend to take active and constructive steps in a global response, without prejudice to sovereignty of States.

I Global Strategy

5. Recognizing that climate change is a global problem of unique character and taking into account the remaining uncertainties in the field of science, economics and response options, we *consider* that a global response, while ensuring sustainable development* of all countries, must be decided and implemented without further delay based on the best available knowledge such as that resulting from the IPCC assessment. Recognizing further that the principle of equity and the common but differentiated responsibility of countries should be the basis of any global response to climate change, developed countries must take the lead. They must all commit themselves to actions to reduce their major contribution to the global net emissions and enter into and strengthen co-operation with developing countries to enable them to adequately address climate change without hindering their national development goals and objectives. Developing countries must, within the limits feasible, taking into account the problems regarding the burden of external debt and their economic circumstances, commit themselves to appropriate action in this regard. To this end, there is a need to meet the requirements of developing countries, that adequate and additional financial resources be mobilized and the best available environmentally sound technologies be transferred expeditiously on a fair and most favourable basis.

* Statement of sustainable development as agreed at the 15th session of UNEP Governing Council (Annex II UNEP/GC 15/L.37)

II Policy Considerations for Action

6. We *reaffirm* that, in order to reduce uncertainties, to increase our ability to predict climate and climate change on a global and regional basis, including early identification of as yet unknown climate-related issues, and to design sound response strategies, there is a need to strengthen national, regional and international research activities in climate, climate change and sea-level rise. We recognize that commitments by governments are essential to sustain and *strengthen* the necessary research and monitoring programmes and the exchange of relevant data and information, with due respect to national sovereignity. We *stress* that special efforts must be directed to the areas of uncertainty as identified by the IPCC. We *maintain* that there is a need to intensify research on the social and economic implications of climate change and response strategies. We *commit* ourselves to promoting the full participation of developing countries in these efforts. We *recognize* the importance of supporting the needs of the World Climate Programme, including contributions to the WMO Special Fund for Climate and Atmospheric Environmental Studies. The magnitude of the problem being addressed is such that no nation can tackle it alone, and we stress the need to strengthen international co-operation. In particular, we *invite* the 11th Congress of the World Meteorological Organization, in the formulation of plans for the future development of the World Climate Programme, to ensure that the necessary arrangements are established in consultation with UNEP, UNESCO (and its IOC), FAO, ICSU and other relevant international organizations for effective co-ordination of climate- and climate-change-related research and monitoring programmes. We *urge* that special attention be given to the economic and social dimensions of climate and climate-change research.
7. In order to achieve sustainable development in all countries and to meet the needs of present and future generations, precautionary measures to meet the climate challenge must anticipate, prevent, attack or minimize the causes of, and mitigate the adverse consequences of, environmental degradation that might result from climate change. Where there are threats of serious or irreversible damage, lack of full scientific certainty should not be used as a reason for

postponing cost-effective measures to prevent such environmental degradation. The measures adopted should take into account different socioeconomic contexts.

8. The potentially serious consequences of climate change, including the risk for survival in low-lying and other small island States and in some low-lying coastal, and arid and semi-arid areas of the world, give sufficient reasons to begin by adopting response strategies even in the face of significant uncertainties. Such response strategies include phasing out the production and use of CFCs, efficiency improvements and conservation in energy supply and use, appropriate measures in the transport sector, sustainable forest management, afforestation schemes, developing contingency plans for dealing with climate-related emergencies, proper land-use planning, adequate coastal-zone management, review of intensive agricultural practices and the use of safe and cleaner energy sources with lower or no emissions of carbon dioxide, methane, nitrous oxide and other greenhouse gases and ozone precursors, paying special attention to new and renewable sources.

   Further actions should be pursued in a phased and flexible manner on the basis of medium- and long-term goals and strategies and at the national, regional or global level, taking advantage of scientific advances and technological developments to meet both environmental and economic objectives.

9. We *note* that *per capita* consumption patterns in certain parts of the world, along with a projected increase in world population, are contributing factors in the projected increase in greenhouse gases.

10. We *agree* that the ultimate global objective should be to stabilize greenhouse gas concentrations at a level that would prevent dangerous anthropogenic interference with climate.

11. We *stress*, as a first step, the need to stabilize, while ensuring sustainable development of the world economy, emissions of greenhouse gases not controlled by the Montreal Protocol on Substances that Deplete the Ozone Layer. Contributions should be equitably differentiated according to countries responsibilities and their level of development. In this context, we acknowledge efforts already undertaken by a number of countries to meet this goal.

12. Taking into account that the developed world is responsible for about three-quarters of all emissions of greenhouse gases, we *welcome* the decisions and commitments undertaken by the European Community with its Member States, Australia, Austria, Canada, Finland, Iceland, Japan, New Zealand, Norway, Sweden, Switzerland, and other developed countries to take actions aimed at stabilizing their emissions of CO_2, or CO_2 and other greenhouse gases not controlled by the Montreal Protocol, by the year 2,000 in general at 1990 level, yet recognizing the differences in approach and in starting point in the formulation of the above targets. We also acknowledge the initiatives of some other developed countries which will have positive effects on limiting emissions of greenhouse gases. We *urge* all developed countries to establish targets and/or feasible national programmes or strategies which will have significant effects on limiting emissions of greenhouse gases not controlled by the Montreal Protocol.

    We *acknowledge*, however, that those developed countries with as yet relatively low energy consumption (measured on a *per capita* or other appropriate basis) which can be reasonably expected to grow, and some countries with economies in transition, may establish targets, programmes and/or strategies that accommodate socioeconomic growth, while improving the energy efficiency of their economic activities.
13. We *urge* developed countries, before the 1992 UN Conference on Environment and Development, to analyse the feasibility of and options for, and, as appropriate in light of these analyses, to develop programmes, strategies and/or targets for a staged approach for achieving reductions of all greenhouse gas emissions not controlled by the Montreal Protocol, including carbon dioxide, methane and nitrous oxide, over the next two decades and beyond.
14. We *recommend* that in the elaboration of response strategies, over time, all greenhouse gases, sources and sinks be considered in the most comprehensive manner possible, and also that limitation and adaptation measures be addressed.
15. We *recognize* that developing countries have as their main priority alleviating poverty and achieving social and economic development and that their net emissions must grow

from their – as yet – relatively low energy consumption to accommodate their development needs. Narrowing the gap between the developed and the developing world would provide a basis for a full partnership of all nations and would assist the developing countries in dealing with the climate change issue. To enable developing countries to meet incremental costs required to take the necessary measures to address climate-change and sea-level rise, consistent with their development needs, we *recommend* that adequate and additional financial resources should be mobilized and best available environmentally sound technologies transferred expeditiously on a fair and most favourable basis. Developing countries also should, within the limits feasible, take action in this regard.

16. The specific difficulties of those countries, particularly developing countries, whose economies are highly dependent on fossil fuel production and exportation, as a consequence of action taken on limiting greenhouse gas emissions, should be taken into account.

17. We *recommend* that consideration should be given to the need for funding facilities, including the proposed World Bank/UNEP/UNDP Global Environmental Facility, a clearing-house mechanism and a new possible international fund composed of adequate additional and timely financial resources and institutional arrangements for developing countries; taking into account existing multilateral and bilateral mechanisms and approaches. Such funding should be related to the implementation of the framework convention on climate change and any other related instruments that might be agreed upon. In the meantime, developed countries are urged to co-operate with developing countries to support immediate action in addressing climate change including sea-level rise without imposing any new conditionality on developing countries.

18. We *recommend* further that resources be assessed. Such assessments, to be conducted as soon as possible, should include country studies and mechanisms to meet the financing needs identified, taking note of the approaches developed under the Montreal Protocol.

19. Financial resources channelled to developing countries should, *inter alia*, be directed to:

(i) Promoting efficient use of energy, development of lower and non-greenhouse-gas-emitting energy technologies and paying special attention to safe and clean new and renewable sources of energy;
(ii) Arranging expeditious transfer of the best available environmentally sound technology on a fair and most favourable basis to developing countries and promoting rapid development of such technology in these countries;
(iii) Co-operating with developing countries to enable their full participation in international meetings on climate change;
(iv) Enhancing atmospheric, oceanic and terrestrial observational networks, particularly in developing countries, to facilitate conducting research, monitoring and assessment of climate change and the impact on those countries;
(v) Rational forest management practices and agricultural techniques which reduce greenhouse gas emissions;
(vi) Enhancing the capacity of developing countries to develop programmes to address climate change, including research and development activities and public awareness and education.

Funding should also be directed to the creation of regional centres to organize information networks on climate change in developing countries.

20. Appropriate economic instruments may offer the potential for achieving environmental improvements in a cost-effective manner. The adoption of any form of economic or regulatory measures would require careful and substantive analyses. We *recommend* that relevant policies make use of economic instruments appropriate to each country's socioeconomic conditions in conjunction with a balanced mix of regulatory approaches.
21. We note that energy production and use account for nearly half of the enhanced radiative forcing resulting from human activities and is projected to increase substantially in the absence of appropriate response actions. We *recognize* the promotion of energy efficiency as the most cost-effective

immediate measure, in many countries, for reducing energy-related emissions of carbon dioxide, methane, nitrous oxide and other greenhouse gases and ozone precursors, while other safe options such as no- or lower-greenhouse-gas-emitting energy sources should also be pursued. These principles apply to all energy sectors. Transport energy use attracts special attention of many of us in the light of its role in many developed countries and of its expected importance in many developing countries.

22. We *recognize* that there is no single quick-fix technological option for limiting greenhouse gas emissions. However, we are *convinced* that technological innovation as well as individual and social behaviour and institutional adaptations is a key element of any long-term strategy that deals with climate change in a way that meets the goal of sustainable development. Therefore, we *urge* all countries, the developed countries in particular, to intensify their efforts and international co-operation in technological research, development and dissemination of appropriate and environmentally sound technologies, including the reassessment and improvement of existing technologies and the introduction of new technologies.

23. We *urge* that environmentally sound and safe technologies be utilized by all sectors in all countries to the fullest extent possible and further *urge* all countries, developed and developing, to identify and take effective measures to remove barriers to the dissemination of such technologies. To this end, the best available environmentally sound and safe technologies should be transferred to developing countries expeditiously on a fair and most favourable basis.

24. We *note* that the conservation of the worlds forests in their role as reservoirs of carbon along with other measures, are of considerable importance for global climatic stability, keeping in mind the important role of forests in the conservation of biological diversity and the protection of soil stability and of the hydrological system. We *recognize* the need to reduce the rate of deforestation in consonance with the objective of sustained yield development and to enhance the potential of the worlds forests through improved management of existing forests and through vigorous programmes of reforestation and afforestation, and to support

financially the developing countries in this regard through enhanced and well-co-ordinated international cooperation including strengthening Tropical Forest Action Plan (TFAP) and International Tropical Timber Organization (ITTO). We *recommend* that the protection and management of boreal, temperate, subtropical and tropical forest ecosystems must be well-co-ordinated and preferably compatible with other possible types of action related to reduction of emission of greenhouse gases, rational utilization of biological resources, provision of financial resources, and the need for more favourable market conditions for timber and timber products. The developing countries should be able to realize increased revenue from these forests and forest products.

25. We also *recognize* that forests and forest products play a key social and economic role in many nations and communities. We *recognize* that States have, in accordance with the Charter of the United Nations and the principles of international law, the sovereign right to exploit their own resources pursuant to their own environmental policies, and the responsibility to ensure that activities within their jurisdiction or control do not cause damage to the environment of other States or of areas beyond the limits of national jurisdiction.

26. We *recommend* that appropriate precautionary and control measures be developed and implemented at regional, subregional and country levels as appropriate to counter the increasing degradation of land, water, genetic and other productive resource bases by drought, desertification and land degradation.

Observatories on climate and climate change and observatories on ecosystems should be encouraged to work together on the consequences of drought. Studies must be undertaken on drought and desertification.

We *stress* that stepped-up financial and scientific contributions should be provided to facilitate these efforts.

27. We *recommend* that similar measures be adopted to address the particular problems and needs, including funding, of low-lying coastal and small vulnerable island countries, some of whose very existence is placed at risk by the consequences of climate change.

III Global Framework Convention on Climate Change

28. We *call* for negotiations on a framework convention on climate change to begin without delay after a decision is taken by the 45th Session of the General Assembly of the United Nations recommending ways, means and modalities for further pursuing these negotiations. Taking note of all the preparatory work, particularly the recommendations adopted on 26 September 1990 by the ad hoc working group of government representatives and regional economic integration organizations to prepare for negotiations on a framework convention on climate change, we *urge* all countries and regional economic integration organizations to join in these negotiations and *recognize* that it is highly desirable that an effective framework convention on climate change, containing appropriate commitments and any related instruments as might be agreed upon on the basis of consensus, be signed in Rio de Janeiro during the United Nations Conference on Environment and Development. We *welcome* the offer of the Government of the United States of America to host the first negotiating meeting.
29. We *recommend* that such negotiations take account of the possible elements compiled by the IPCC, and that the framework convention on climate change be framed in such a way as to gain the support of the largest possible number of countries while allowing timely action to be taken. We *reaffirm* our wish that this convention contain real commitments by the international community. We *stress*, given the complex and multifaceted nature of the problem of climate change, the need for new and innovative solutions, including the need to meet the special needs of developing countries.
30. We also *welcome* the invitations of Thailand and Italy to host workshops, respectively on the feasibility of forestry options, and on all technologies for energy production and use and their transfer to developing countries.
31. We *believe* that a well-informed public is essential for addressing and coping with as complex an issue as climate change, and the resultant sea-level rise, and *urge* countries, in particular, to promote the active participation

at the national and, when appropriate, regional levels of all sectors of the population in addressing climate-change issues and developing appropriate responses. We also *urge* relevant United Nations organizations and programmes to disseminate relevant information with a view to encouraging as wide a participation as posssible.

Appendix III

Mountains and the Cryospere – Proposals for the World Conservation Strategy*

Issues

Mountains cover approximately 40 per cent of the earth's surface and 10 per cent of the world's population live in these regions, including some of the poorest tribal and indigenous peoples. Many more people living on the plains depend on the mountains for water, energy, forests, recreation and spiritual needs. Many of the world's endangered species, including valuable medicinal plants and cultivars, are located in specialized mountain microhabitats.

The cryosphere, the snow and ice regions of the world, may extend at the height of the northern winter to 60 per cent of the earth's surface. The cryosphere plays a vital role in regulating climate change through the albedo effect, whereby heat is reflected back to space.

Mountains and the cryosphere are currently under great threat. In Europe more than 50 per cent of the forests are dying prematurely because of pollution and dryness. The melting of Alpine glaciers is contributing to sea-level rises, whilst the melting of the permafrost has led to disastrous rockfalls and damage to roads and buildings. Human activities in this fragile ecosystem, particularly tourism, are causing damage, especially in wetlands. Erosion, the result of deforestation and poor grassland management, is becoming more frequent, often producing disastrous flooding downstream. Poverty is intense in regions such as the Himalayas, Andes, Atlas and Ethopian highlands. Traditional cultures and ways of life are threatened. There are many health problems in mountain regions, respiratory ailments and diarrhoeas

* Draft statement presented to the General Assembly of the World Conservation Union (IUCN): Perth, December 1990

in the tropics and cancers, a consequence of ozone depletion, in the north.

Recommendations

1. That an international campaign for mountain reforestation be established. Climate changes will probably result in the potential for forest planting at higher altitudes and latitudes. Mountain forests may thus make a major contribution in the carbon sink. Appropriate indigenous species planting should be encouraged.
2. Governments and NGOs should be urged to increase the areas of mountains and cryosphere under protection. Permanent ice (including Antarctica) might be declared a wilderness park with minimal human intervention. Below the nival zones a major part of the land area might be protected. Protection should extend to the cultural as well as natural heritage. Provision should be made for the safeguarding of traditional agriculture, lifestyles and identities. Consideration should be given to declaring mountains car-free zones. Avalanche and erosion control should be included in protection.
3. An international campaign should be established to promote better grassland management in mountain areas. Appropriate species, especially those that are acid-resistant, should be encouraged.
4. Alternative energy sources should be proposed for mountain areas including solar hydrogen, minihydrels, biomass and wind.
5. Alternative employment schemes are necessary for depressed mountain areas. These should emphasise self-reliance and take full advantage of appropriate technology, including communications (telecottaging).
6. Research on biodiversity in mountain areas should be continued with red databooks for all major ranges.
7. An internationally co-ordinated education, training and awareness programme should be established, drawing together existing efforts.
8. Templates for future conservation and development should be prepared for all mountain ranges involving the fullest participation of local communities.

9. A network should be created of existing mountain centres interested in mountain conservation. This network should have access to a common databank of relevant information, resources, consultant skills, etc.
10. An international fund should be established for the protection of the mountains with an international mandate and a small secretariat. This fund should draw not only on the UN but on NGOs and the business world.

Appendix IV

IPCC 3: Report of the Response Strategies Working Group*

The Scenarios

A. The 2030 High Emissions Scenario

- Few or no steps are taken to reduce emissions.
- Montreal Protocol comes into effect but without strengthening and with less than 100 per cent compliance by countries.
- Fossil fuels continue to dominate energy supply.
- Costs of solar electricity remain at current levels.

B. The 2060 Low Emissions Scenario

- Energy efficiency improves more rapidly due to such factors as technology transfer.
- The share of primary energy provided by natural gas increases significantly.
- Full compliance with the Montreal Protocol is realized.
- Tropical deforestation is halted and a global reforestation effort begins.

C. The Control Policies Scenario

- Rapid penetration of renewable energy sources in the last half of the next century (solar electricity costs fall to half of current levels).
- Increased utilization of nuclear power.
- Montreal Protocol is strengthened to include a full phase-out of CFCs and freezes on methyl-chloroform and carbon tetrachloride.

---

* Extracted from *Climate Change – A Reader's Guide to the IPCC Reports* (London: Greenpeace, 1990).

- Agricultural policies yield reductions in emissions.

D. The Accelerated Policies Scenarios

are similar to the Control Policies Scenario but feature much more rapid development and penetration of renewable energy sources (partly encouraged by global adoption of carbon fees of up to $50/tonne of coal by the year 2025) plus greater use of nuclear power.

Only in the Alternative Accelerated Policies scenarios is the goal of the 1988 Toronto Conference "The Changing Atmosphere", for a 20 per cent reduction in CO_2 emissions by 2005, achieved.

Measures to Reduce Greenhouse Gases

In the shopping list of options produced by RSWG which could reduce greenhouse gas emissions are:

- energy efficiency and conservation improvements;
- fuel substitution by energy sources which have lower or no greenhouse gas emissions, e.g. move from coal to gas, or to non-carbon fuels such as solar or nuclear;
- reduction of emissions by removal, recirculation or fixation;
- management and behavioural changes (e.g. increased work in homes through information technology) and structural changes (e.g. shift towards public transport);
- phasing out CFC emissions and careful assessments of the greenhouse gas potential of proposed substitutes;
- sustainable forest management and afforestation;
- review of agricultural practices to improve productivity and reduce wastes.

Carbon Reductions in the Energy Sectors

1. Transportation Sector

Options include fuel substitution via fuels derived from biomass, e.g. ethanol; electric or hydrogen-fuelled vehicles from non-carbon-based electricity sources; and fuel-efficiency improvements. All three options were regarded as more expensive than current options, because petroleum fuels are relatively cheap.

2. Buildings Sector

"Energy requirements in new homes could be roughly half of the current average for new homes . . . and the technical potential for retrofits could average 25 per cent." "Reductions of energy use in existing commercial buildings by at least 50 per cent may be technical feasible, and new commercial buildings could be up to 75 per cent more efficient."

3. Industry Sector

"The technical potential for efficiency improvements in the industry sector ranges from 13 per cent in some subsectors to over 40 per cent in others." "Considerable opportunities for energy savings also exist in the industrial sector by the recycling of energy-intensive waste."

4. Electricity Sector

"The technical potential for greater efficiency in generation is in the range of 15–20 per cent." "Overall, efficiency of the electricity system can be improved through the use of least cost utility planning."

Note: Research for Greenpeace has demonstrated larger technical potential improvements in efficiency than those indicated here.

Feasibility and Costs of Reducing Energy Emissions

Feasibility

Around twenty case studies were provided by individual countries, the European Community and the International Energy Agency. Though developed using widely different dates, assumptions and policy measures (where the effects of these were included), the EIS subgroup made some general conclusions, including:

- reducing the annual growth in CO_2 emissions in developing countries from over 3 per cent to 2 per cent;
- slowing the growth in East European countries and the USSR to stabilize emissions over the next two decades;

- stabilizing or reducing West European emissions by early in the next decade;
- slowing the growth in CO_2 emissions in North American and Pacific OECD countries.

Costs

Some preliminary cost figures were provided which suggested:

(a) "significant emissions reductions are available at low or negative costs when tested at social discount rates." "By 2020 this might amount to around 20 per cent of global emissions."

(b) "A significant further tranche of emission abatement is potentially available at relatively moderate resource costs."

(c) "as the scale of abatement rises, marginal abatement costs will escalate."

Note: Many of the case studies presented included extremely high "business as usual" emissions, way above those produced in RSWGs "High Emissions" scenario. The UK case study, for example, has been criticized by a range of academics and organizations for this. A wide range of alternative scenarios have been produced worldwide showing that significant carbon reductions are available at low or zero cost. For example, a study by the Batelle Memorial Institute on behalf of the US Environmental Protection Agency covered eight OECD and East European countries. It concluded that most countries "appear to be capable of holding emissions constant at no real cost to their economies", and achieving 20 per cent reductions at costs ranging from negative to 0.75 per cent of GNP in 2020.

Policy Measures

The measures fall generally into three groups:

- Information measures.
- Economic measures, e.g carbon taxes, pollution permits.
- Regulatory measures, e.g. efficiency standards and least-cost planning.

REPORT OF THE SUBGROUP ON AGRICULTURE, FORESTRY, AND OTHER HUMAN ACTIVITIES

Future Greenhouse Gas Emissions

Future greenhouse gas emissions in these sectors are difficult to predict because of many uncertainties, but the IPCC estimates that CO_2 emissions from deforestation could range between 1.1 and 3.9 billion tonnes of carbon in 2020 (an increase of 40 to 90 per cent), and that methane emissions from flooded rice will increase by around 30 to 80 per cent in 2025.

Policies for Reducing Emissions

Near-Term Options

The opportunities for reducing greenhouse gas emissions and enhancing carbon sinks in the near term include:

1. Forestry

- Improve fuel efficiency, e.g. woodstoves.
- Introduce sustainable harvesting and natural forest.
- Large-scale afforestation and forest-protection plans.
- Ensure the health of existing forests by reducing pollution, e.g. acid deposition.
- User available agricultural land for forestry.

2. Agriculture

- Reduce biomass burning.
- Reduce methane emissions through management of livestock wastes and expansion of supplemental feeding practices for livestock.
- Reduce nitrous oxide emissions through improved fertilizers, and through better use of animal manure.

3 Waste Management

- Use landfill gas-collection systems and flaring.
- Use biogas systems to treat wastewater in development countries.

Longer-Term Options

4. Forestry

- Implement improved global forestry management.
- Increase wood production and forest productivity.

5. Agriculture

- Management of water regimes, development of new crop systems, and more efficient use of fertilizers could lead to a 10 to 30 per cent reduction in methane emissions from flooded rice cultivation.
- Reduce methane emissions from livestock by up to 25–75 per cent through improved stock systems.
- Reduce trace gas emissions from biomass burning, land conversion and crops through adoption of sustainable agriculture, optimizing fertilizer use and improved pasture management and grazing systems.

International Options

6. Forestry

- Strengthen and extend the Tropical Forestry Action Plan, the role of the International Timber Trade Organization (ITTO) and the role of development banks, the International Monetary Fund, United Nations agencies and other multilateral or bilateral international organizations in helping developing countries to achieve conservation and the sustainable development of forests and agriculture.
- At a special workshop on tropical forest in São Paulo, Brazil, in January 1990 it was "recommended that a World Forest Conservation Protocol, covering temperate, boreal and tropical forests, be developed in the context of a climate convention process that also addresses energy supply and use."

Planting Trees to Soak up Carbon Emissions

The amount of afforestation required to balance total anthropogenic carbon dioxide emissions would be about 1 billion hectares (ha), or an area the size of Europe from the Atlantic to the Urals. Currently, only one million hectares per year are being afforested. Afforestation on such a scale is impracticable; hence forestry measures can play a minor but none the less important contributory role in reducing the build-up of greenhouse gases. It is estimated that in 20–50 years, very roughly, up to 200 million hectares might be planted, which would require an average rate of 4–10 million ha per year. Such an area could store up to about 20 per cent of present carbon dioxide emissions for a limited time.

Afforestation Scenarios: three scenarios of replanting were evaluated by IPCC, including planting 6 million ha/year over ten years, 89 million ha/year over twenty years, or 10 million ha/year over thirty years to offset 5 per cent, 13 per cent or 26 per cent of the 5.5GtC currently released from burning fossil fuels annually. The estimated costs total $222.4, $3.0 or $4.0 billion per year, respectively.

Note: The major tree-planting schemes implied by the above would require huge agroforestry projects. There is major doubt as to the productivity of such schemes over time, and uncertainty as to the quantities of carbon which would actually be locked up in the trees. Turning the trees into paper or burning the wood releases carbon back into the atmosphere.

REPORT OF THE COASTAL ZONE MANAGEMENT SUB GROUP

Responses to sea-level rise

The responses required to protect human life and property from sea-level rise and associated impacts fall into three categories: Retreat, Accommodation and Protection.

Retreat – involves no effort to protect the land from the sea.

Accommodation – implies that people continue to use the land at risk but do not attempt to prevent it from being flooded.

Protection – involves hard structures such as sea walls and dykes, as well as soft solutions such as dunes and vegetation, to protect the land from the sea so that existing land uses can continue.

Environmental Implications of Policy Options

Two-thirds of the world's fish catch, and many marine species, depend on coastal wetlands for their survival. Without human interference (i.e. the "retreat" option), ecosystems could migrate landward as sea level rises and remain largely intact, although the total area of wetlands would decline. Under the "protection" option, a much larger proportion of these ecosystems would be lost, especially if hard structures block their landward migration.

Economic Implications of Policy Options

No response strategy can completely eliminate the economic impacts of climate change. In the "retreat" options, coastal landowners and communities would suffer from loss of property, resettlement costs, and the costs of rebuilding infrastructure. Under "accommodation", there would be changing property values, increasing damage from storms, and costs for modifying infrastructures. Under the "protection" option, nations and communities would face the costs for the necessary structures.

"It is urgent for coastal nations to begin the process of adapting to sea-level rise – not because there is an impending catastrophe, but because there are opportunities to avoid adverse impacts by action now, opportunities that may be lost if the process is delayed."

Recommendations

National

- By the year 2,000, coastal nations should implement comprehensive coastal-zone management plans, and identify coastal areas at risk.
- Nations should ensure that coastal development does not increase vulnerability to sea-level rise.
- Emergency preparedness and coastal-zone response mechanisms need to be reviewed and strengthened.

International Co-operation

- A continuing international focus on the impacts of sea-level rise needs to be maintained.
- Technical assistance for developing nations should be provided and co-operation stimulated.

- International organizations should support national efforts to limit population growth in coastal areas.

Research, Data and Information

- Research on the impacts of global climate change on sea-level rise should be strengthened.
- A global ocean observing network should be developed and implemented.
- Data and information on sea-level change and adaptive options should be made widely available.

Appendix V

Extract from Second World Climate Conference – Conference Statement*

Part I: Main Conclusions and Recommendations

A. *Greenhouse Gases and Climate Change*

1. Emissions resulting from human activities are substantially increasing atmospheric concentrations of the greenhouse gases. These increases will enhance the natural greenhouse effect, resulting on average in an additional warming of the earth's surface. The Conference agreed that this and other scientific conclusions set out by the IPCC reflect the international consensus of scientific understanding of climate change. Without actions to reduce emissions, global warming is predicted to reach 2 to 5° C over the next century, a rate of change unprecedented in the past 10,000 years. The warming is expected to be accompanied by a sea-level rise of 65 cm + 35 cm by the end of the next century. There remain uncertainties in predictions, particularly in regard to the timing, magnitude and regional patterns of climate change.

2. Climate change and sea-level rise would seriously threaten low-lying islands and coastal zones. Water resources, agriculture and agricultural trade, especially in arid and semi-arid regions, forests, and fisheries are especially vulnerable to climate change. Climate change may compound existing serious problems of the global mismatch between resources, population and consumption. In many cases the impacts will be felt most severely in regions already under stress, mainly in developing countries.

3. Global warming induced by increased greenhouse gas concentrations is delayed by the oceans; hence, much of the change is still to come. Inertia in the climate system due to the influence

* Extracted from SWCC 1990, Document No. 13 (mimeo)

of the oceans, the biosphere and the long residence times of some greenhouse gases means that climate changes that occur may persist for centuries.

4. Natural sources and sinks of greenhouse gases are sensitive to a change in climate. Although many of the response or feedback processes are poorly understood, it appears likely that as climate warms, these feedbacks will lead to an overall increase rather than a decrease in greenhouse gas concentrations.

5. The historical growth in emissions has been a direct consequence of the increase of human population, rising incomes, the related exploitation of fossil fuels by industrialized societies and the expansion of agriculture. Under "business as usual" assumptions it is projected that emissions will continue to grow in the future as a consequence of a projected doubling of energy consumption in the first half of the twenty-first century and an expected doubling of population by the latter half. As a result, the effect of human-induced greenhouse gas concentrations on the earths radiation balance would correspond to a doubling of carbon dioxide by 2025 unless remedial action is taken.

6. Over the last decade, emissions of carbon dioxide (CO_2) contributed 55 per cent of the increased radiative forcing produced by greenhouse gases from human activities. The CFCs contributed about 24 per cent of the past decades changes and methane 15 per cent, with the balance due to other greenhouse gases. With controls on CFCs under the Montreal Protocol, the relative importance of CO_2 emissions will increase, provided the substitutes for CFCs have minimal greenhouse warming potential. Some 75 per cent of total CO_2 emissions have come from the industrialized countries.

7. The above emissions can be expected to change the planets atmosphere and climate, and a clear scientific consensus has been reached on the range of changes to be expected. Although this range is large, it is prudent to exercise, as a precautionary measure, actions to manage the risk of undesirable climate change. In order to stabilize atmospheric carbon dioxide concentrations by the middle of the twenty-first century at about 50 per cent above pre-industrial concentrations, a continous worldwide reduction of net carbon dioxide emissions by 1 to 2 per cent per year, starting now, would be required. The Intergovernmental Panel on Climate change (IPCC) also considered three other emissions scenarios, which would not lead to stabilization of CO_2 concentrations in

the twenty-first century. A 15–20 per cent reduction in methane emissions would stabilize atmospheric concentrations of that gas.

8. This Conference concludes that technically feasible and cost-effective opportunities exist to reduce CO_2 emissions in all countries. Such opportunities for emissions reductions are sufficient to allow many industrialized countries to stabilize CO_2 emissions from the energy sector and to reduce these emissions by at least 20 percent by 2005. The measures include increasing the efficiency of energy use and employing alternative fuels and energy sources. As additional measures to achieve further cost-effective reductions are identified and implemented, even greater decreases in emissions would be achieved in the subsequent decades. In addition, reversing the current net losses in forests would increase storage of carbon. The economic and social costs and benefits of such measures should be urgently examined by all nations. An internationally co-ordinated assessment should be undertaken through the IPCC.

9. Countries are urged to take immediate actions to control the risks of climate change with initial emphasis on actions that would be economically and socially beneficial for other reasons as well. Nations should launch negotiations on a convention on climate change and related legal instruments without delay and with the aim of signing such a convention in 1992.

B. Use of Climate Information in Assisting Sustainable Social and Economic Development

Climate data, analyses, and eventually climate predictions can contribute substantially to enhancing the efficiency and security of economic and developmental activities in environmentally sustainable ways. These benefits are particularly important in food and wood production, water management, transportation, energy planning and production (including assessment of potential resources of biomass, hydropower, solar and wind energy), urban planning and design, human health and safety, combating of drought and land degradation, and tourism. This requires both data on the climate system, and its effective application. Data acquisition, collection, management and analysis must be more vigorously supported in all countries, and special assistance must be provided to developing countries through international co-operation. Transfer of techniques for applying climate information should be accelerated through more widespread use of software (e.g. CLICOM) for readily available personal computers and other means. Further development of methods for predicting short-term variations in climate and the environmental and social impacts should be vigorously pursued. These advances would provide enormous economic and other welfare benefits in coping with droughts, prolonged rain, and periods of severe hot and cold weather. Such predictions will require major steps forward in ocean-atmosphere-biosphere observing systems. Much greater efforts are also needed to increase involvement in these fields by developing countries, especially through increased education and training.

C. Priorities for Enhanced Research and Observational Systems

1. A consensus exists among scientists as summarized in the Report of Working Group 1 of the IPCC that climate change will occur due to increasing greenhouse gases. However, there is substantial scientific uncertainty in the details of projections of future climate change. Projections of future regional climate and climate impacts are much less certain than those on a global scale. These uncertainties can be narrowed only through research addressing the following priority areas:

clouds and the hydrological cycle
greenhouse gases and the global carbon and biogeochemical cycles
oceans: physical, chemical and biological aspects; and exchanges with the atmosphere
palaeoclimatic studies
polar ice sheets and sea ice
terrestrial ecosystems.

2. These subjects are being addressed by national programmes, the World Climate Research Programme and the International Geosphere-Biosphere Programme and other related international programmes. Increased national support and substantially increased funding of these programmes is required if progress on the necessary time scale is to be made in reducing the uncertainties.

3. Present observational systems for monitoring the climate system are inadequate for operational and research purposes. They are deteriorating in both industrialized and developing regions. Of special concern is the inadequacy of observation systems in large parts of the southern hemisphere.
4. High priority must be placed on the provision and international exchange of high-quality, long-term data for climate-related studies. Data should be available at no more than the cost of reproduction and distribution. A full and open exchange of global and other datasets needed for climate-related studies is required.
5. There is an urgent need to create a Global Climate Observing System (GCOS) built upon the World Weather Watch Global Observing System and the Integrated Global Ocean Service System and including both space-based and surface-based observing components. GCOS should also include the data communications and other infrastructure necessary to support operational climate forecasting.
6. GCOS should be designed to meet the needs for:

(a) climate system monitoring, climate change detection and response monitoring, especially in terrestrial ecosystems
(b) data for application to national economic development, and
(c) research towards improved understanding, modelling and prediction of the climate system.

7. The main components of such a GCOS would be:

(1) an improved World Weather Watch Programme;
(2) the establishment of a global ocean observing system (GOOS) of physical, chemical and biological measurements;
(3) the maintenance and enhancement of monitoring programmes of other key components of the climate system, such as the distribution of important atmospheric constituents (including the Global Atmosphere Watch), changes in terrestrial ecosystems, clouds and the hydrological cycle, the earth's radiation budget, ice sheets, and precipitation over the oceans.

8. The further development and implementation of the GCOS concept should be pursued, with urgency, by scientists, governments and international organizations. All countries must ensure a full and open exchange of the datasets needed for climate system research, process and impact studies, and modelling.

9. The impacts of climate variability on human socioeconomic systems have provided major constraints to development. Climate change may compound these constraints. In semi-arid regions of Africa, drought episodes have been directly responsible for major human disasters. Research undertaken during the first decade of the WCP and through other international and national programmes has improved drought early-warning systems, including FAOs Global Early Warning System, and increased the reliability of climate impact analyses. But much more remains to be done. Intensified efforts are required to refine further our ability to predict short-term climate variability, anticipate climate impacts, and identify rational strategies to mitigate or prevent adverse effects. The threat of climate change brings new challenges to the future well-being of people. This requires greater efforts to understand impacts of climate change. Mitigation and adaptation strategies are also essential. Immediate steps to be taken include:

(a) national and regional analyses of the impacts of climate variability and change on society, and study of the range of response and adaptation options available;
(b) closer co-operation and communication among natural and social scientists, to ensure that climate considerations are accounted for in development planning;
(c) significant increases in resources to carry out impact/adaptation studies.

10. Improvements in energy efficiency and non-fossil-fuel energy technologies are of paramount importance, not only to reduce greenhouse gas emissions but to move to more sustainable development pathways. Such advances will require research and development, as well as technology transfer and co-development.

11. A specific initiative would create a network of regional, interdisciplinary research centres, located primarily in developing countries and focusing on all the natural science, engineering and social science disciplines required to support fully integrated studies of global change and its impacts. These centres would conduct research and training on all aspects of global change and study the interaction of regional and global policies.

D. Public Information

People need better information on the crucial role climate plays in development and the additional risks posed by climate change. Governments, intergovernmental and non-governmental organizations should give more emphasis to providing accurate public information on climate issues. The public information and education and training component in the WCP and IGBP must also be expanded.

Appendix VI

Possible Elements for Inclusion in a Framework Convention on Climate Change*

Preamble

In keeping with common treaty practice, including the format of the Vienna Convention, the Climate Change Convention would contain a preamble which might seek to address some or all of the following items:

- a description of the problem and reasons for action (need for timely and effective response without awaiting absolute scientific certainty);
- reference to relevant international legal instruments (such as the Vienna Convention and Montreal Protocol) and declarations (such as UNGA Resolution 43/53 and Principle 21 of the Stockholm Declaration);
- recognition that climate change is a common concern of humankind, affects humanity as whole, and should be approached within a global framework, without prejudice to the sovereignty of states over the air space superadjacent to their territory as recognized under international law;
- recognition of the need for an environment of a quality that permits a life of dignity and well-being for present and future generations;
- reference to the balance between the sovereign right of states to exploit natural resources and their concomitant duty to protect and conserve climate for the benefit of humankind, in a manner not to diminish either;
- endorsement and elaboration of the concept of sustainable development;

---

* Extracted from IPCC

- recognition of the need to improve scientific knowledge (e.g. through systematic observation) and to study the social and economic impacts of climate change, respecting national sovereignty;
- recognition of the importance of the development and transfer of technology and of the circumstances and needs, particularly financial, of developing countries; need for regulatory, supportive and adjustment measures to take into account different levels of development and thus differing needs of countries;
- recognition of the responsibility of all countries to make efforts at the national, regional and global levels to limit or reduce greenhouse gas emissions and prevent activities which could adversely affect climate, while bearing in mind that:
  - most emissions affecting the atmosphere at present originate in industrialized countries where the scope for change is greatest;
  - implementation may take place in different time frames for different categories of countries and may be qualified by the means at the disposal of individual countries and their scientific and technical capabilities;
  - emissions from developing countries are growing and may need to grow in order to meet their development requirements and thus, over time, are likely to represent an increasingly significant percentage of global emissions;
- recognition of the need to develop strategies to absorb greenhouse gases, i.e. protect and increase greenhouse gas sinks; to limit or reduce anthropogenic greenhouse gas emissions; and to adapt human activities to the impacts of climate change.

Other key issues which will have to be addressed during the development of the preambular language include:

- Should humankind's interest in a viable environment be characterized as a fundamental right?
- Is there an entitlement not to be subjected, directly or indirectly, to the adverse effects of climate change?
- Should there be a reference to the precautionary principle?
- In view of the interrelationship among all greenhouse gases, their source and sinks, should they be treated collectively?

- Should countries be permitted to meet their aggregate global climate objectives through joint arrangements?
- Should reference be made to weather-modification agreements such as the ENMOD Treaty as relevant legal instruments?
- Is there a common interest of humankind in the development and application of technologies to protect and preserve climate?
- Does the concept of sustainable development exclude or include the imposition of new conditionality in the provision of financial assistance to developing countries, and does it imply a link between the protection and preservation of the environment, including climate change, and economic development, so that both are to be secured in a coherent and consistent manner?
- Should the preamble address the particular problems of countries with an agricultural system vulnerable to climate change and with limited access to capital and technologies, recognizing the link with sustainable development?
- Is there a minimum standard of living which is prerequisite to adopting response strategies to address climate change?

Definitions

As is the practice, definitions will need to be elaborated in a specific article on definitions. The terms which will need to be defined will depend on the purpose of the Convention and thus the language used by the negotiating parties.

General Obligations

Following the format of such treaties as the Vienna Convention, an article would set out the general obligations agreed to by the parties to the Convention. Such obligations may relate to, for example:

- the adoption of appropriate measures to protect against the adverse effects of climate change, to limit, reduce, adapt to and, as far as possible, prevent climate change in accordance with the means at the disposal of individual countries and their scientific and technical capabilities; and to avoid creating other environmental problems in taking such measures;

- the protection, stabilization and improvement of the composition of the atmosphere in order to conserve climate for the benefit of present and future generations;
- taking steps having the effect of limiting climate change but already justified on other grounds;
- the use of climate for peaceful purposes only, in a spirit of good-neighbourliness;
- co-operation by means of research, systematic observation and information exchange in order to understand better and assess the effects of human activities on the climate and the potential adverse environmental and socioeconomic impacts that could result from climate change, respecting national sovereignty;
- the encouragement of the development and transfer of relevant technologies, as well as the provision of technical and financial assistance, taking into account the particular needs of developing countries to enable them to fulfil their obligations;
- co-operation in the formulation and harmonization of policies and strategies directed at limiting, reducing, adapting to and, as far as possible, preventing climate change;
- co-operation in the adoption of appropriate legal or administrative measures to address climate change;
- provision for bilateral, multilateral and regional agreements or arrangements not incompatible with the Convention and any annex/protocol, including opportunities for groups of countries to fulfil the requirements on a regional or subregional basis;
- co-operation with competent international organizations effectively to meet the objectives of the Convention;
- the encouragement of and co-operation in the promotion of public education and awareness of the environmental and socioeconomic impacts of greenhouse gas emissions and of climate change;
- the strengthening or modification if necessary of existing legal and institutional instruments and arrangements relating to climate change;
- a provision on funding mechanisms.

Other key issues which will have to be addressed in the process of elaborating this article include the following:

- Should there be a provision setting any specific goals with respect to levels of emissions (global or national) or atmospheric concentrations of greenhouse gases while ensuring stable development of the world economy, particularly stabilization by industrialized countries, as a first step, and later reduction of CO_2 emissions and emissions of other greenhouse gases not controlled by the Montreal Protocol? Such provisions would not exclude the application of more stringent national or regional emissions goals than those which may be provided for in the Convention and/or any annex/protocol.
- In the light of the preambular language, should there be a provision recognizing that implementation of obligations may take place in different time frames for different categories of country and/or may be qualified by the means at the disposal of individual countries and their scientific and technical capabilities?
- Should there be a commitment to formulate appropriate measures such as annexes, protocols or other legal instruments, and if so, should such formulation be on a sound scientific basis or on the basis of the best available scientific knowledge?
- In addressing the transfer of technology particularly to development countries, what should be the terms of such transfers (i.e. commercial vs. non-commercial, preferential vs. non-preferential, the relationship between transfers and the protection of intellectual property rights)?
- Should funding mechanisms be limited to making full use of existing mechanisms or also entail new and additional resources and mechanisms?
- Should provision be made for environmental impact assessments of planned activities that are likely to cause significant climate change as well as for prior notice of such activities?
- What should be the basis of emission goals, e.g. total emission levels, *per capita* emissions, emissions per GNP, emissions per energy use, climatic conditions, past performance, geographic characteristics, fossil fuel resource base, carbon

intensity per unit of energy, energy intensity per GNP, socioeconomic costs and benefits, or other equitable considerations?

- Should the particular problem of sea-level rise be specifically addressed?
- Is there a link between nuclear stockpiles and climate change?

Institutions

It has been the general practice under international environmental agreements to establish various institutional mechanisms. The parties to a Climate Change Convention might, therefore, wish to make provision for a Conference of the Parties, an Executive Organ and a Secretariat.

The Conference of the Parties may, among other things: keep under continuous review the implementation of the Convention and take appropriate decisions to this end; review current scientific information; and promote harmonization of policies and strategies directed at limiting, reducing, adapting to and, as far as possible, preventing climate change.

Questions that will arise in developing provisions for appropriate institutional mechanisms include:

- Should any of the Conventions institutions (e.g. the Conference of the Parties and/or the Executive Organ) have the ability to take decisions *inter alia* on response strategies or functions in respect of surveillance, verification and compliance that would be binding on all the parties and, if so, should such an institution represent all of the parties or be composed of a limited number of parties, e.g. based on equitable geographic representation?
- What should be the role of the Secretariat?
- What should be the decision-making procedures, including voting requirements (eg. consensus, majority)?
- If a trust fund or other financial mechanism were established under the Convention, how should it be administered?
- Should specific and/or other bodies be established on a permanent or *ad hoc* basis, to provide advice and make recommendations to the Conference of the Parties concerning

research activities and measures to deal with climate change?
- Should the composition of the above bodies reflect equitable climatic or geographic representation?
- Should there be a provision for working groups, e.g. on scientific matters as well as on socioeconomic impacts and response strategies?
- Is there a need for innovative approaches to institutional mechanisms in the light of the nature of the climate change-issue?
- What should be the role of non-governmental organizations?

Research, Systematic Observations and Analysis

It would appear to follow general practice to include provision for co-operation in research and systematic monitoring. In terms of research each party might be called upon to undertake, initiate, and/or co-operate in, directly or through international bodies, the conduct of research on and analysis of:

- physical and chemical processes that may affect climate;
- substance, practices, processes and activities that could modify the climate;
- techniques for monitoring and measuring greenhouse gas emission rates and their uptake by sinks;
- improved climate models, particularly for regional climates;
- environmental, social and economic effects that could result from modifications of climate;
- alternative substances, technologies and practices;
- environmental, social and economic effects of response strategies;
- human activities affecting climate;
- coastal areas with particular reference to sea-level rise;
- water resources; and
- energy efficiency.

The parties might also be called upon to co-operate in establishing and improving, directly or through competent international bodies, and taking fully into account national legislation and relevant ongoing activities at the national, regional and international levels, joint or complementary programmes for systematic monitoring and analysis of climate, including a possible worldwide system; and co-operate in ensuring the collection, validation and transmission of research, observational data and analysis through appropriate data centres.

Other issues that will arise in developing this provision include:

- should consideration be given to the establishment of panels of experts or of an independent scientific board responsible for the co-ordination of data collection from the above areas of research and analysis and for periodic assessment of the data?
- should provision be made for on-site inspection?
- should there be provision for open and non-discriminatory access to meteorological data developed by all countries?
- should a specific research fund be established?

Information Exchange and Reporting

Precedents would suggest the inclusion of a provision for the transmission of information through the secretariat to the Conference of the Parties on measures adopted by them in implementation of the Convention and of protocols to which they are party. In an annex to the Vienna Convention, the types of information exchanged are specified and include scientific, technical, socioeconomic, commercial and legal information.

For the purposes of elaborating this provision, issues having to be addressed by the negotiating parties include the following:

- Is there a need for the elaboration of a comprehensive international research programme in order to facilitate co-operation in the exchange of scientific, technological and other information on climate change?
- Should parties be obliged to report on measures they have adopted for the implementation of the Convention, with the possible inclusion of regular reporting on a comparable basis of their emissions of greenhouse gases?
- Should each party additionally be called upon to develop a national inventory of emissions, strategies and available technologies for addressing climate change? If so, the Con-

vention might also call for the exchange of information on such inventories, strategies and technologies.

Development and Transfer of Technology

While the issue of technology has been addressed in the section on General Obligations, it might be considered desirable to include separate provisions on technology transfer and technical co-operation. Such provisions could call upon the parties to promote the development and transfer of technology and technical co-operation, taking into account particularly the needs of developing countries, to enable them to take measures to protect against the adverse effects of climate change, to limit, reduce and, as far as possible, prevent climate change, or to adapt to it.

Another issue which will arise is: should special terms be attached to climate-related transfers of technology (such as a preferential and/or non-commercial basis and assured access to, and transfer of, environmentally sound technologies on favourable terms to developing countries), taking into consideration the protection of intellectual property rights?

Settlement of Disputes

It would be usual international practice to include a provision on the settlement of disputes that may arise concerning the interpretation or application of the Convention and/or any annex/protocol. Provisions similar to those in the Vienna Convention for the Protection of the Ozone Layer might be employed, i.e. voluntary resort to arbitration or the International Court of Justice (with a binding award) or, if neither of those options is elected, mandatory resort to conciliation (with a recommendatory award).

Other Provisions

It would be the usual international practice to include clauses on the following topics:

- amendment of the Convention;
- status, adoption and amendment of annexes;
- adoption and entry into force of, and amendments to, protocols;
- signature;
- ratification;
- accession;
- right to vote;
- relationship between the Convention and any protocol(s);
- entry into force;
- reservations;
- withdrawal;
- depository;
- authentic texts.

Annexes and Protocols

The negotiating parties may wish the Convention to provide for the possibility of annexes and/or protocols. Annexes might be concluded as integral parts of the Convention, while protocols might be concluded subsequently (as in the case of the Montreal Protocol to the Vienna Convention on Protection of the Ozone Layer). While it is recognized that the Convention is to be all-encompassing, the negotiating parties will have to decide whether greenhouse gases, their sources and sinks, are to be dealt with individually, in groups, or comprehensively, in annexes or protocols to the Convention. The following, among others, might also be considered as possible subjects for annexes or protocols to the Convention:

- agricultural practices;
- forest management;
- funding mechanisms;
- research and systematic observations;
- energy conservation and alternative sources of energy;
- liability and compensation;
- international emissions trading;
- international taxation system;
- development and transfer of climate-change-related technologies.

Issues that will arise in connection with the development of annexes and protocols include:

- Timing, i.e. negotiating parties advocating a more action-orientated Convention may seek to include specific obligations in annexes as opposed to subsequent protocols and/or negotiate one or more protocols in parallel with the Convention negotiations.
- Sequence, i.e. if there is to be a series of protocols, in what order should they be taken up?

Appendix VII

The Bellerive Initiative on Alpine Action

Prince Sadruddin Aga Khan proposed the following ten priority points for immediate action at the conclusion of the Conference held in Geneva on 28 February 1989;

1. An urgent review of current knowledge and future scenarios in key areas: e.g. forests, water, agriculture, tourism, community development, etc.
2. The preparation of an Alpine Conservation Strategy, including a Code of Ethics.
3. The preparation of a Master Plan for implementing degrees of protection for ecologically sensitive areas.
4. The support of existing preparations for an Alpine Convention rationalizing existing legislation and exploring new legal instruments.
5. The establishment of a network of scientific monitoring centres with a common database and connected by on-line computers.
6. The promotion of printed and audiovisual materials for educational and public information purposes.
7. The convening of regular conferences assessing the state of the Alps organized by a focal institution and network.
8. The promotion of more effective local community management, including the provision of independent technical advice on the environmental and economic implications of development activities.
9. The encouragement of the financial community to contribute to conservation.
10. The creation of a fund to support an action group to animate these activities.

Source: Pitt, D. and Stone, P. *Towards Alpine Action* (Geneva: Bellerive, 1989)

Index

Alp Action

The Alps, still perceived as an unspoilt paradise, are being affected by pollution largely exported from the industrial regions of Europe: one cannot place a dome over the Alps! Yet they are one of the natural wonders of the Earth, the source of Europe's vital hydrological system – four major European rivers, the Rhône, the Rhine, the Po and the Danube originate in the Alps. Although scientists disagree on the extent and causes of the environmental degradation, there exists a general consensus on the problems threatening the mountains – in particular deforestation, erosion, pollution and climate change, the decline of local communities, their agriculture and traditions.

The current situation in the Alps reflects almost every developmental and environmental problem currently encountered in the world. Alp Action was launched by the Bellerive Foundation as part of the need to adopt a positive approach to the environment. The fund is a real and mutually trusting partnership between the corporate sector, the scientific and conservation communities, and promotes projects designed to preserve the Alpine environment.

Taking into account the diversity of threats facing the Alps, Alp Action was convinced, from the start, that only a large number of small, targeted projects would collectively respond to the urgency of the situation. Plenty of practical solutions are proposed – reafforestation, the protection of species of flora and fauna and their habitats, studies of the impacts of infrastructure in Alpine regions, the stimulation of an authentic Alpine culture, education and awareness campaigns. Activities such as soft tourism, for instance, can be beneficial to the Alpine environment. In fact, it is difficult to label any particular activity as abusive. Rather, it is the way in which managers plan and implement tourist policies that counts. The key to successful Alpine management is in long-term thinking rather than short-term profit. Decentralization, small-scale technology and the philosophy of quality not quantity is a healthier basis for economic growth.

Alp Action acts as a broker between business and environmental organizations. Its role is to identify sponsors wishing to finance selected projects that are part of the Alp Action portfolio, to co-ordinate their implementation and to manage the media aspects linked to these projects' achievements. The sponsors are closely associated with the projects through information to the media. This media feedback, and the sponsors' participation in positive environmental action, often gets full support from the sponsors' personnel. Finally, and most important, Alp Action partnerships are the basis of international awareness campaigns, combining the marketing efforts of companies with Alp Action's educational programmes.

The International Institute for Applied Systems Analysis

is a nongovernmental research institution, bringing together scientists from around the world to work on problems of common concern. Situated in Laxenburg, Austria, IIASA was founded in October 1972 by the academies of science and equivalent organizations of twelve countries. Its founders gave IIASA a unique position outside national, disciplinary, and institutional boundaries so that it might take the broadest possible view in pursuing its objectives:

To promote international cooperation in solving problems arising from social, economic, technological, and environmental change

To create a network of institutions in the national member organization countries and elsewhere for joint scientific research

To develop and formalize systems analysis and the sciences contributing to it, and promote the use of analytical techniques needed to evaluate and address complex problems

To inform policy advisors and decision makers about the potential application of the Institute's work to such problems

The Institute now has national member organizations in the following countries:

Austria
The Austrian Academy of Sciences

Bulgaria
The National Committee for Applied Systems Analysis and Management

Canada
The Canadian Committe for IIASA

Czech and Slovak Federal Republic
The Committee for IIASA of the Czech and Slovak Federal Republic

Finland
The Finnish Committee for IIASA

France
The French Association for the Development of Systems Analysis

Germany
Association for the Advancement of IIASA

Hungary
The Hungarian Committee for Applied Systems Analysis

Italy
The National Research Council (CNR)
and the National Commission for Nuclear and Alternative Energy Sources (ENEA)

Japan
The Japan Committee for IIASA

Netherlands
The Netherlands Organization for Scientific Research (NWO)

Poland
The Polish Academy of Sciences

Sweden
The Swedish Council for Planning and Coordination of Research (FRN)

Union of Soviet Socialist Republics
The Academy of Sciences of the Union of Soviet Socialist Republics

United States of America
The American Academy of Arts and Sciences

For Product Safety Concerns and Information please contact our EU representative GPSR@taylorandfrancis.com
Taylor & Francis Verlag GmbH, Kaufingerstraße 24, 80331 München, Germany

www.ingramcontent.com/pod-product-compliance
Ingram Content Group UK Ltd.
Pitfield, Milton Keynes, MK11 3LW, UK
UKHW011011180926
475701UK00004B/675
9780415849630